EXPLORING THE MYSTERIES OF LIMESTONE

with Hydroplate Theory

ISBN 979-8-9868637-3-3
Printed in USA.
ejm.basementworkshop@gmail.com

Retailers can order this book through Ingramcontent.com

Limestone is so common in the eastern U.S. that it is crushed to make inexpensive gravel.

WHAT IS LIMESTONE?

Limestone is any type of sedimentary rock whose chemistry includes the chemical compound calcium carbonate, $CaCO_3$. About 20 percent of all sedimentary rocks fall into the category of limestone. Some types of limestone are from organic sources, and others are inorganic. Sometimes other elements, like magnesium or iron, can be mixed into limestone, producing rocks such as dolomite, travertine or colored marbles.

For a map showing all the areas of the world that have some type of limestone, see page 36.

TYPES OF LIMESTONE

The most common type of limestone is a gray, fine-grained, non-fossiliferous rock (shown above as gravel and as boulders) and doesn't have a special name like other forms of limestone do. Quarries that mine this type of limestone are known simply as ***limestone quarries***. The quarry shown here, at right, located in Central Pennsylvania, crushes its limestone into large gravel for making roads, into small gravel for making concrete, and into powder for making cement and concrete. Some quarries have a higher grade of limestone that can be cut into blocks called "dimension stone," which are suitable for making buildings or monuments.

Occasionally, fine-grained limestone can be split into flat "sheets" which are useful for a type of printing known as ***lithography***. ("Litho" means "stone.") The flat surfaces of these rocks are highly polished, then designs are drawn onto them. Ink is then rolled onto the surface of the stone and paper is pressed onto the ink to make a print. The print is called a lithograph.

This limestone is sheet-like, but not good enough for printing.

In certain areas of the world, fine-grained limestone contains many fossils. One of the most famous ***fossiliferous limestone*** sites is the Solnhofen Quarry in southern Germany. The limestone here contains an abundance of fossils—plants, soft-bodied invertebrates, insects, birds, and mammals. The most famous fossil from this site is the archaeopteryx. Many marine animals are also found here including exquisitely preserved jellyfish, crustaceans, and crinoids (sea lilies). The degree of preservation is astounding, which gives us a hint about the conditions in which these fossils formed.

White "Carrara" marble is from massive quarries in Tuscany, Italy.

If limestone is put under extreme heat and/or pressure, the calcium carbonate molecules can be forced to rearrange themselves into small crystals. When the crystals become large enough that they start to be visible, the limestone has turned into ***marble.*** Marble is classified as a ***metamorphic rock*** because it has changed. ("Meta" means "change," and "morph" means "shape.") The crystals are what give marble its slightly sparkly luster. The colors in some marbles are due to tiny amounts of minerals such as iron oxide (red and yellow) or feldspar (pink or blue).

Marble is the most popular type of rock for making fine art sculptures, like the Lincoln Memorial in Washington, D.C. It has also been used to make stone buildings, including the famous Taj Mahal in India.

The crystals found in marble are called ***calcite*** crystals. Calcite is pure $CaCO_3$. Rarely, calcite crystals will grow very large, making spectacular specimens for mineral collectors. (Notice the penny next to the calcite crystal shown below.) The shape that calcite crystals form is called a ***rhombohedron***, which some people describe as a "squished box" shape. No matter what size the crystal is, it will always have this rhombohedral shape. Extremely pure calcite crystals can be almost transparent. More often, they are semi-transparent, like the one shown with the penny. When calcite contains small amounts of other minerals such as iron, copper or cobalt, it can be light shades of orange, green, blue or brown. Occasionally, gray limestone will have streaks or clumps of pure white calcite.

rhombohedral calcite crystal

colored calcites

gray limestone with a layer of calcite

When $CaCO_3$ molecules form crystals, they will usually arrange themselves into rhombohedral calcite, but if they are put into an extremely high-pressure situation, they can organize themselves into an ***orthorhombic*** shape. This crystal shape might be described as a normal, rectangular box, not a squished one. When this arrangement of the $CaCO_3$ molecules occurs, we call the crystal ***aragonite***, named after a place where it was found in 1797: Molina de Aragón, in Spain. Specimens of aragonite rarely look like they have box-shaped crystals, however. They can form spheres, with long "needles" coming from the center. They can also make stalactite formations in limestone caves. Organic forms of aragonite are found in the shells of many marine mollusks.

Calcite often has other chemical elements mixed into it, but one particular element—magnesium—is found in calcite so often that when it is present in sufficient quantities, we call the rock "***dolomite***" instead of limestone. (Dolomite was named after an 18th century French geologist named de Dolomieu.) This mountain range in northern Italy is made of dolomite. The formula for dolomite is: $CaMg(CO_3)_2$. This means that the CO_3 has no preference for whether it hangs onto a Ca atom or an Mg atom. Molecules of $CaCO_3$ are mixed with molecules of $MgCO_3$. The reason that magnesium can easily sneak in and replace calcium is that Ca and Mg have the same electrical charge: (2+). The significance of dolomite will be discussed later in the book.

Another place where you find additional elements mixed into calcite is around hot springs. Boiling water comes up from deep in the earth and brings dissolved minerals with it. As the water cools, the minerals come out of the solution and collect around the vents. At Mammoth Hot Springs in Yellowstone Park, shown below, calcite precipitates out of the hot water, forming strange and beautiful mineral formations. The yellow and brown coloring is probably due to the presence of iron or sulfur. This type of calcite is called ***travertine***.

Calcite can also precipitate out of cool water. Limestone formations found in cool water are called ***tufa***. Tufa "towers" are found in a handful of lakes around the world, including Pyramid Lake in Nevada, USA, shown in the photo below. Cave formations can also be classified as tufa formations. As rainwater drips down through limestone rock, it dissolves some of the rock and picks up calcium and carbonate molecules. When the water reaches a cave, it drips from the ceiling. As the water evaporates, calcium atoms and carbonate molecules come together to form calcium carbonate. These solid particles clump together and form crusty mineral deposits. If the water drips from one location long enough, the mineral deposits will form long ***stalactites***. The time it takes for a stalactite to form depends on how much water is coming into the cave and the amount of dissolved minerals that the water is carrying.

Pyramid Lake, Nevada

When calcium carbonate particles form in seawater, there aren't any cave ceilings to stick to. Seawater does often have tiny solid particles floating in it, however, such as sand grains or small pieces of seashell. To a molecule of calcium carbonate, these tiny particles seem huge, so the molecules are glad to stick to them. As the tiny particles tumble around in the water they can become evenly coated with calcium carbonate, forming a spherical shape that we call an ***oöid*** (*oh-oyd),* Greek for "egg-like." This photo was taken with a microscope and shows ooids that were cut in half to reveal their internal structure. The red scale bar shows that they range from about .5 to 1.5 millimeters.

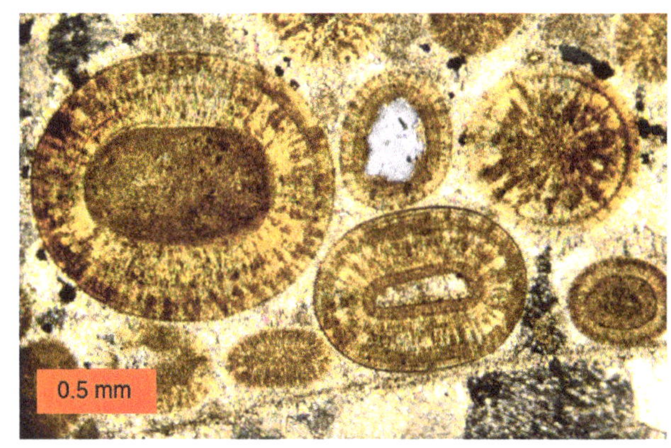

As the oöids collect more $CaCO_3$ molecules, they become too heavy to stay suspended in the water and will fall to the seafloor. Over time, the oöids can become permanently stuck together and form ***oölitic limestone*** *(oh-oh-lit-ik)*. Some oölitic limestones have formed recently and have many tiny organic forms stuck in them, like the 3-mm sea star shown in the photo on the left. If the oölitic limestone has a very large number of shells in it, like the photo on the right, it can be called ***coquina limestone***, from the Spanish word for seashell. Coquina limestone is considered to be an ***organic*** form of limestone, meaning that living things played a large roll in its formation. Mollusks (clams, whelks, conchs, snails, oysters, etc.) and corals take calcium carbonate out of the water and use it to make their shells (or "skeletons" in the case of corals). The presence of mollusk shells in these rocks indicates that the water that the mollusks were living in had a plentiful supply of dissolved $CaCO_3$ for building shells.

oölitic limestone, close up

coquina limestone

The rock shown here is a type of oölitic limestone that doesn't have any shells in it. The oöids are very small, so, overall, the rock looks fairly smooth—not too different from ordinary limestone. This type of limestone was likely made entirely by chemical processes, with the oöids forming around inorganic particles such as sand grains. Because the grain size is so small and the amount of $CaCO_3$ available for cementing the particles was so high, this type of oolitic limestone is ideal for making buildings and monuments. The Soldiers and Sailors monument in Indianapolis was made from oölitic limestone taken from quarries in Indiana.

fine-grained oölitic limestone

The strangest type of limestone is ***chalk***. The most famous chalk formation is the White Cliffs of Dover, on the southern coast of England. A matching set of cliffs is found on the other side of Channel, on the coast of France. The soft, white rock in these cliffs used to be mined and sold as chalk for writing on blackboards. (Today, this natural chalk is being conserved, so "school chalk" is made of a mineral called gypsum, $CaSO_4$.) The rock under the water in the channel is also made of chalk.

Natural chalk is made of shells from microscopic animals with strange names like ***foraminiferans*** *(for-AM-in-IF-er-ans)* and ***coccolithophores*** *(co-co-LITH-o-fores)*. These microscopic animals are classified as protozoans (single-celled organisms), often falling into the sub-category of algae. We tend to think of algae as being mushy green stuff, so this classification may seem strange. These shelled algae certainly are bizarre in shape and structure. The coccolithophore shown here is made of a living cell that uses calcium carbonate to make plates that it wears almost like armor. When the living cell dies, the plates fall off and eventually drift to the bottom of the sea. These algal animals grow quickly, and are famous for producing "algal blooms" where the water becomes thick with them in a very short amount of time. When a large volume of these calcium-rich shells pile up, they can form a chalk deposit.

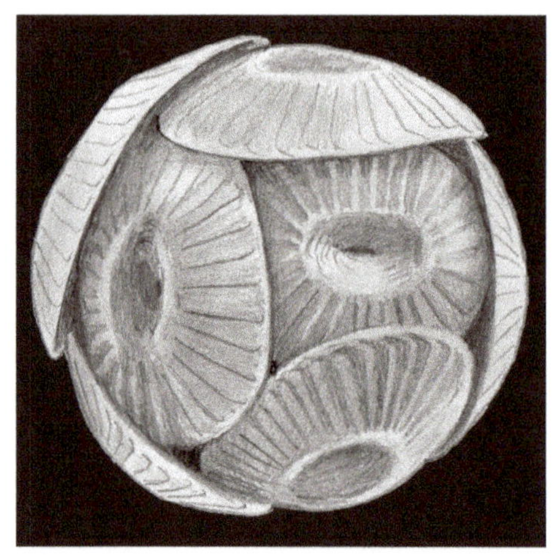

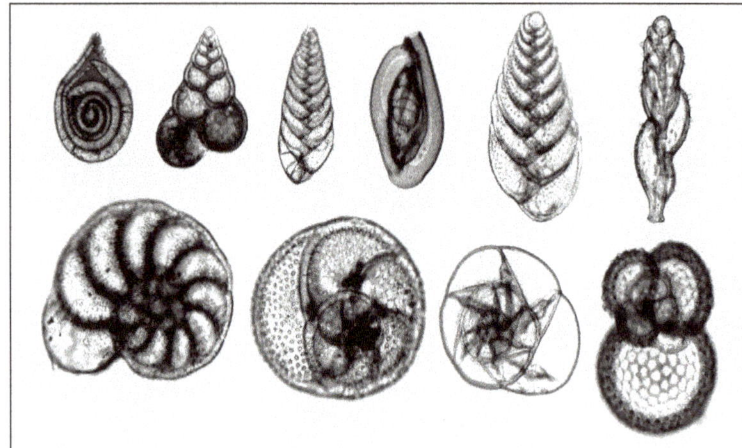

shells of foraminiferans

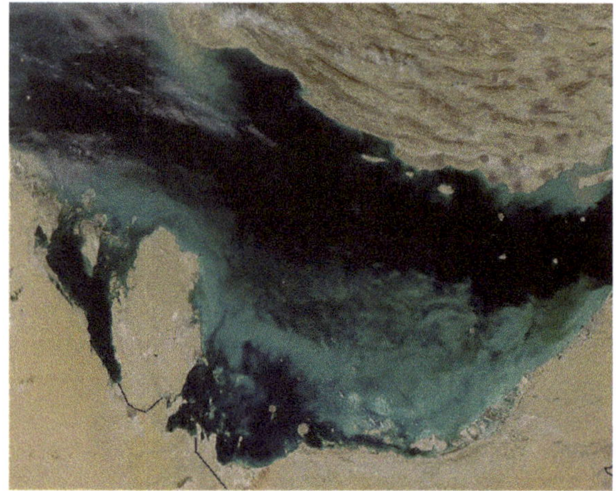

The light green is an algal bloom in the Persian Gulf.

LIMESTONE CHEMISTRY

Limestone isn't always easy to identify by sight, but a simple chemical test can reveal whether a rock contains caclium carbonate. A few drops of hydrochloric acid will make limestone "fizz," producing bubbles of carbon dioxide. In some cases, even vinegar will work, but the reaction will not be as energetic. The chemical reaction that produces the bubbles is as follows:

$$2\,HCl + CaCO_3 \rightarrow CaCl_2 + CO_2 + H_2O$$

The hydrochloric acid, HCl, releases hydrogen ions, H^+, which break apart the $CaCO_3$. When all the atoms are allowed to recombine into new molecules, the result is calcium chloride, carbon dioxide, and water. Other types of rock will not fizz when hydrochloric acid is dropped onto them. Only calcium carbonate (or possibly magnesium carbonate, $MgCO_3$) will produce CO_2 bubbles.

There is another reaction that involves both calcium carbonate and carbon dioxide, and this reaction will help us understand one of the mysteries of limestone.

$$H_2O + CO_2 + CaCO_3 \leftrightarrow Ca^{2+} + 2(HCO_3^{-1})$$

Notice that the arrow is pointing in both directions. This means that the reaction can go either way, left to right, or right to left. We can start with water, carbon dioxide (dissolved into the water) and calcium carbonate and end up with a solution that contains calcium ions (atoms that carry an electrical charge of positive 2), and two molecules of bicarbonate. When the reaction goes the other way, from right to left, we start with a solution that contains both calcium and carbonic acid, and produce a molecule of calcium carbonate along with a molecule of carbon dioxide and a molecule of water. If we figure in the masses of the atoms, this equation tells us that for every 100 grams of limestone that is formed, 44 grams of carbon dioxide will be released into the environment.

Another interesting feature of limestone chemistry is that whereas most substances tend to be more soluble as the temperature rises, limestone becomes less soluble. This is why you see mineral "scales" forming on the inside of hot water pipes and tea kettles. Technically, calcium carbonate is insoluble in water at any temperature. This is why limestone can be used to make buildings and monuments. Rainwater won't dissolve the stone. However, our chemical equation tells us that if carbon dioxide is present in the water, the situation changes. Unfortunately for old buildings, as CO_2 in the atmosphere increases, it is reacting with water droplets in the upper atmosphere to create acid rain. When rain droplets contain dissolved CO_2, our chemical equations tells us what will happen to the $CaCO_3$ in the limestone blocks.

HOW MOST GEOLOGY TEXTBOOKS EXPLAIN LIMESTONE

As we've seen, there are quite a few types of rock that fall into the category of limestone. Some have a direct connection to marine creatures that make shells from calcium carbonate. We see organic limestone being formed today, with the broken shells of marine creatures being part of the recipe. We also observe chemical processes forming carbonate rocks such as travertine. Geologists acknowledge that both biological and chemical processes can produce limestones. However, when they try to explain vast areas of limestone, such as the limestone layers in the Grand Canyon, or entire mountain ranges made of dolomite or limestone, they usually choose the biological explanation. The choice of biology over chemistry seems strange when we consider how large many of these formation are. Yet, the textbooks authors dutifully write the approved storyline, saying that a layer of limestone means that this area was once a "warm, shallow sea." The sea has to be warm and shallow because these conditions will favor the life cycle of shelled mollusks and corals, the evaporation of water, and the precipitation of limestone. Over time, they say, the marine creatures living in these warm, shallow seas died, and their shells (or skeletons in the case of corals) gradually disintegrated and then were cemented together to make stone. Each limestone layer of the Grand Canyon was made by a warm, shallow sea, with vast amounts of time occurring between each layer. The seafloor apparently sank at just the right rate to keep the shallow sea from getting too shallow as the limestone built up on the bottom. The sandstone layers between the limestone layers are explained as desert eras.

Many of the layers in the Grand Canyon are made of limestone. The most famous limestone layer is the Redwall Limestone, which ranges from 150 meters to 240 meters thick.

The Rock of Gibraltar is made of limestone.

These weathered limestone formations are in China.

FOUR LIMESTONE MYSTERIES

The first limestone mystery is related to limestone chemistry. We learned that for every molecule of $CaCO_3$ that is formed, a molecule of carbon dioxide is also formed. Chemists have calculated that for every 100 grams of $CaCO_3$ produced, 44 grams of CO_2 are released. Can we figure out approximately how many grams of limestone are in the crust of the earth, and then calculate how much CO_2 would have been produced?

The most efficient way to do this estimate is to focus on the element carbon, the "C" in $CaCO_3$. How much carbon is locked up in limestone? A researcher published his estimate in the journal *Science* in the year 2000. He said that there are about 6,000,000 x 10^{16} grams of carbon in the world's limestone. If we wrote out the number instead of using exponent notation, it would look like this: 60,000,000,000,000,000,000,000.

How does this number compare to other sources of carbon (and potentially carbon dioxide)?

ATMOSPHERE: **72** x 10^{16}
PLANTS and ANIMALS: **200** x 10^{16}
COAL, OIL and GAS: **413** x 10^{16}
OCEANS: **3740** x 10^{16}
LIMESTONE: **6,000,000** x 10^{16}

Even if this amount of CO_2 accumulated gradually, over supposed millions of years, it would still be a problem because CO_2 tends to stay "in circulation," being passed from the oceans to the atmosphere, to plants, to animals, back to the atmosphere, back to plants, etc. To get CO_2 out of circulation, it must somehow get "locked into" stone (or coal or oil that never gets used as fuel).

Limestone has far more carbon (and potential CO_2) than all other sources combined! If limestone formed on the surface of the earth, so that the resulting CO_2 was released into the air and water, the volume of CO_2 would have been enough to poison the planet hundreds of times over. Plants do use carbon dioxide, but they can only work so fast to remove it from the atmosphere, and it is toxic to animal life. Plants are able to deal with 72 x 10^{16} grams, not 6,000,000 x 10^{16} grams. If limestone could not have formed at earth's surface, is there an alternative explanation?

The second limestone mystery is best seen in the Grand Canyon, where the cliffs show us rock layers that would otherwise have been hidden deep underground. These layers are remarkably flat. Each layer is amazingly uniform in thickness, and can be hundreds of miles wide. Limestone layers alternate with layers of shale and sandstone. Shale and sandstone are not believed to be products of a warm, shallow sea, so those limestone-producing seas would have to have completely disappeared for long intervals of time while these other layers formed. Then, suddenly it was back to a "warm, shallow sea" again. We are expected to believe that during these millions years the layers escaped normal erosional forces and maintained their flat boundaries. Were there no violent storms during all those eons? No meteor impacts? No dirt accumulation? No trees? No coral reef structures? Was there no erosion of the top layer of sediment when the waters of the next warm shallow sea began to wash into the region? How did these layers of limestone (and shale and sandstone) remain so flat over the eons, especially when you consider that geologists also expect us to believe that the Colorado River cut the entire canyon in only 6 million years?

A related mystery is found between two particular layers of limestone. Underneath the famous Redwall Limestone is a layer called the Temple Butte Limestone, and under that is the Muav Limestone. The Muav layer is said to have been completed by about 490 million years ago (mya), at the end of the Cambrian period. The Temple Butte layer is said to be from the Upper Devonian time period, running from 382 mya to 358 mya. Subtract 382 from 490, and you find that there are over 100 million years of missing sediments. That's a very long time for nothing to happen. The top of the Muav limestone is flat. How can it have escaped erosion for 100 million years?

Additionally, it is quite a stretch to imagine that the 120-meter-thick Redwall limestone was able to be deposited by a warm, shallow sea imagined to be only 10-20 meters deep. How could such a shallow sea possibly form a layer of rock 6-10 times its depth? For this to happen, we need to assume that the ground underneath the sea was subsiding at exactly the same rate that the limestone was forming. What are the chances of that happening? And what would be the mechanism for the subsidence? Also, we have the "red" in the Redwall limestone, which comes from iron oxide. Where did all the iron come from and when did it stain the limestone?

The third mystery is the worldwide distribution of limestone. Today, organic limestone forms at locations that are between 30 degrees north and 30 degrees south of the equator—places where the water is warm and provides a good environment for shelled marine creatures. However, we find vast areas of limestone and dolostone on all continents, and at all latitudes. Even northern Siberia has great quantities of limestone. Landscapes built on limestone rock are called ***karst*** regions. Karst landscapes often include caves, sinkholes, or strange "sculpture" formations on the surface. If the limestone is deep, it can also provide natural aquifers that store fresh water. Karst is found at all latitudes, not just in the tropics.

This limestone structure was made by ancient people who lived in a high-latitude karst region.

The fourth mystery is why there is such a difference between old limestone and modern limestone. Here is a chart summarizing the differences. (Sources are listed in the bibliography.)

Grand Canyon limestone	modern limestone ("limemud")
$CaCO_3$ looks like calcite	$CaCO_3$ looks like aragonite
grain size is 1-4 microns	grain size is about 20 microns
quartz sand grains mixed in	no quartz sand grains
fossils pointing in same general direction	fossils not in same direction
fossils of delicate, soft-bodied animals	no fossils of soft animals

BONUS MYSTERY: Some limestone contains fossilized jellyfish. Why were they preserved instead of rotting or being eaten by scavengers?

The presence of sand grains in the Canyon limestone indicates that the ancient water was not the "calm, shallow sea" necessary for the slow accumulation of limemud. Sand is not part of limestone chemistry, so it would have washed in from somewhere else. (Coincidentally, above and below some limestone layers there are very thick layers of sandstone.) Why such a difference between old and new limestone?

THE DOLOMITE MYSTERY

We've already learned what dolomite is. Its chemical formula is $CaMg(CaCO_3)_2$. This means that the $CaCO_3$ can be hanging on to either a calcium atom or a magnesium atom. Although dolomitic limestone is a relatively common rock, it is rarely observed forming in current sedimentary environments. For this reason, it is believed that dolomitic rock (also called "dolostone") is formed when limestone is modified by groundwater that is high in magnesium. The dolostone can have more magnesium than calcium, or the level of magnesium can be very low. High-magnesium dolostone is harder than regular limestone, so it is often preferred for making blocks for stone buildings. Dolostone is also less vulnerable to acid rain.

a sample of the mineral dolomite

The problem that dolomite (dolostone) presents is that we can't use uniformitarian assumptions to explain it. ***Uniformitarianism*** says that "the present is the key to the past." This assumption underlies most geological theories. With dolostone, this is certainly not true. Present processes are not producing dolostone. The groundwater theory (where magnesium-rich water comes in after the limestone has formed) has issues. How could groundwater trickle through so much rock so evenly? Large deposits of dolostone are fairly uniform in their chemistry—they are not "patchy" as you would expect from a trickling water source. Also, many dolostone deposits are large, like the Dolomite Mountains in northern Italy. The groundwater theory doesn't seem plausible for these mountains, even if the trickling happened before the mountains rose. This is a vast amount of rock. The photo doesn't show the whole range.

Another reason that the groundwater theory was proposed is that shelled marine organisms, which are invoked as the main mechanism for production of limestone, don't use magnesium to make their shells. A few minor species do, but the vast majority do not. For standard geology, this rules out the possibility that dolostone formed at the same time, and in the same manner, as limestone. It seems that many dolostone formations, like these mountains, remain unexplained. This is the "dolomite problem."

A POSSIBLE SOLUTION TO THESE MYSTERIES

It seems as though the guiding principle of modern geology, uniformitarianism, may not be very helpful in trying to figure out the origin of limestone. As we've seen, current processes are unable to provide good explanations for old limestone rock. What if we set aside this principle and entertain the possibility that the geological history of the world has not been uniform? After all, the people who formulated this uniformitarian idea were not eyewitnesses to the past. The only claim ever made by any human with respect to being an eyewitness to geological events of the distant past (before the pyramids were built) occurs primarily in a text scoffed at by geologists: the writings of the ancient Hebrews, in their book of Genesis. Another equally old version of this account is the ancient Sumerian text called the Enuma Elish, but the Hebrew version provides us with more technical details. Let's go off on a "rabbit trail" that will eventually bring us back to the topic of limestone.

In the book of Genesis, a man claims to have witnessed the crust of the earth break apart (Gen. 7:11) followed by torrential rains that flooded the entire surface of the earth. This account was preserved over many centuries by a culture that highly valued keeping accurate historical records The Israelite scribes were "detail oriented" to the extreme. They did not shy away from recording history as it happened, even if it involved painful accounts of personal or corporate sins. Successive generations faithfully copied these accounts century after century. The Dead Sea scrolls helped to confirm the amazing accuracy of the copies.

There isn't a rational, non-biased reason to automatically discount this recorded history simply because it is old. If ancient peoples really did see an amazing cataclysmic event, and they wanted to let people in the future know about it, what would they have done? Like all ancient peoples, they would have handed down the story as an oral tradition until writing was developed. As the population grew and people groups began to scatter across the globe, they would have taken this oral legend with them, so it is not surprising that we find flood stories in about 200 ancient cultures. The stories are not identical, but they all have at their core an earth-shattering water event that devastated the world and left few survivors. Some of the legends have talking animals or other fanciful details that were obviously added by creative story tellers.

A colorful depiction of Noah and his sons from a 16th century Turkish manuscript. The ark is depicted as a contemporary Turkish sailing vessel, with added windows.

The Hebrew version of the story is by far the longest, the most detailed, and the most realistic. It records an exact day on which the tragedy began, the number of days that rain fell, the depth of the water over the tallest mountain peaks, the number of days the waters prevailed after the rain stopped, and how long it took until the ground was dry. It also tells how a handful of people were able to survive, and gives details about the size and shape of the vessel and what materials were used to build it. The text doesn't read like mythology.

If we take these texts at face value and let them suggest that ancient humans really did witness a global flooding event, the first question that comes to mind is whether a global flood is physically possible. The water cycle as we know it today is not capable of producing enough water to flood the entire globe. The water that pours down as rain comes from evaporation of surface water in oceans, lakes, rivers and puddles. Even if all the surface water ended up in the clouds as rain, the end result would simply be refilling the basins from which the water came, with no net increase in the amount of surface water. To have enough water to cover the earth, we'd need to find a source of water outside the water cycle, and it would help if we could also find a way to reduce the height of our mountains. Amazingly enough, the account we read in Genesis does indeed suggest a source of water that is not part of the water cycle. Here are the opening verses of the flood story found in Genesis. (New International Version.)

11 In the six hundredth year of Noah's life, on the seventeenth day of the second month—on that day all the fountains of the great deep burst forth, and the floodgates of the heavens were opened. 12 And rain fell on the earth forty days and forty nights.

18 The waters rose and increased greatly on the earth, and the ark floated on the surface of the water. 20 The waters rose and covered the mountains to a depth of more than fifteen cubits. 21 Every living thing that moved on land perished—birds, livestock, wild animals, all the creatures that swarm over the earth, and all mankind.

24 The waters flooded the earth for a hundred and fifty days.

The ancient narrator tells us that the very first thing that happened was the bursting open of the "fountains of the great deep." The plain and obvious meaning of the text is that a great volume of water suddenly and violently came bursting out of the ground. The only reason to think anything else is if you are trying to reconcile this verse with other ideas you may have about geological history. Some would like to make the fountains into magma plumes. However, ancient people knew water when they saw it and if we let the text speak to us, instead of trying to read things into it, it clearly portrays a scene where great volumes of water came up from below, immediately followed by torrential rains. Another reason to reject the magma idea is that magma erupting up through the ocean floor would not be able generate massive fountains of water. The ocean is not a closed hydraulic system, and the water would be free to go sideways instead of up. Much energy would be lost as water moved away in every direction. Also, we need to consider the weight of several miles of water pressing down on the ocean floor. Scientists have observed many magma eruptions on the ocean floor and none of them have produced anything even close to vertical jets of water. The magma might produce steam bubbles that rise to the surface, but no jets. The laws of fluid mechanics and our actual observations of underwater magma eruptions both testify against the "fountains of magma" idea.

The next event in the ancient narrative is the commencement of heavy rain. The Hebrew word used here for rain is "geshem," which means a very heavy downpour, not an ordinary rain shower. If the "fountains of the great deep" were, indeed, jets of water, and they were energetic enough to put water high up into the atmosphere for several weeks, they could easily be the source of the forty days of rain. With a credible explanation found for the unusual rainfall, we now need to consider what the "great deep" might have been. Is it plausible to hypothesize a sizable amount of water deep underground?

Until recently, geologists scoffed at the idea that a great volume of water could have come up from below the earth's crust. However, since about 2010, headlines like these have appeared in science journals:

"Ultra Rare Diamond Suggests Earth's Mantle Has an Ocean's Worth of Water." (Scientific American, Sept. 2022)

"New Evidence for Oceans of Water Deep in the Earth" (Brookhaven National Lab, June 2014)

"There's as much water in the Earth's mantle as in all the oceans" (New Scientist, June 2017)

"Found! Hidden Ocean Locked Up Deep in Earth's Mantle" (LiveScience, June 2014)

These headlines are a bit misleading—the water they are talking about isn't liquid H_2O. The "water" is actually molecules of OH^- that are firmly attached to minerals in mantle rock. Often, the mineral cited is "ringwoodite," a form of olivine, Mg,Fe(SiO_4), that can incorporate OH's into its chemical formula. If these OH's could break free from the minerals and then pick up a hydrogen atom that was in the area, they could turn themselves into water molecules. Voila—water from rock!

an artificial sample of ringwoodite

By Jasperox -Photomicrograph taken at University of Hawaii, Previously published: https://www.flickr.com/photos/joerox/4061008289/in/set-72157608270817196, CC BY 3.0, https://commons.wikimedia.org/w/index.php?curid=19689605

Our information about the presence of OH's in mantle rock doesn't come from direct observation, since the location is believed to be 500-600 km below the surface—far deeper than we can drill. Most of what is taught about ringwoodite and its ability to absorb hydroxide ions (OH's) comes from mathematical calculations or computer simulations of mantle conditions, and from observations of artificial samples of ringwoodite produced in labs. Until recently, the only actual sample of naturally occurring ringwoodite came from a meteorite. Yes, one of those things that comes from outer space. Then, in 2014, it was announced that a sample of ringwoodite had been found in a rock of terrestrial origin. A tiny, rough diamond had been found in Brazil, and inside that diamond was an almost microscopic (.1 mm) sample of ringwoodite. This tiny sample, the size of a pin prick, spawned all those grandiose headlines about an ocean of water deep inside the earth. However, because of these headlines, it is no longer considered crazy talk to suggest that there might have been water under the crust of the earth.

Additional support for the possibility of subterranean water came from another unexpected source: the planet Mars. Astronomers have been able to see erosional features on the surface of Mars that look like they could have been created by water. Many scholarly articles have been written about these erosional features, and how they might have been formed by water that gushed up suddenly from beneath Mars's rocky crust. Some speculate that perhaps Mars's ice caps also formed during this event. Most astronomers have now come to terms with the idea that Mars might have had large amounts of water under its crust long ago. The problem of how and why Mars had subcrustal water has been put aside. Not all astronomers agree with the flooding hypothesis. Some have suggested that these erosional features look more like patterns created by intense lightning strikes. Many of Mars's features are very strange, indeed, and it is hard to see how water could have made them. The important point for us is that if a cataclysmic flooding event on Mars merits serious discussion, it is certainly not unreasonable to propose a cataclysmic flooding event on earth, where 75% of the surface is water.

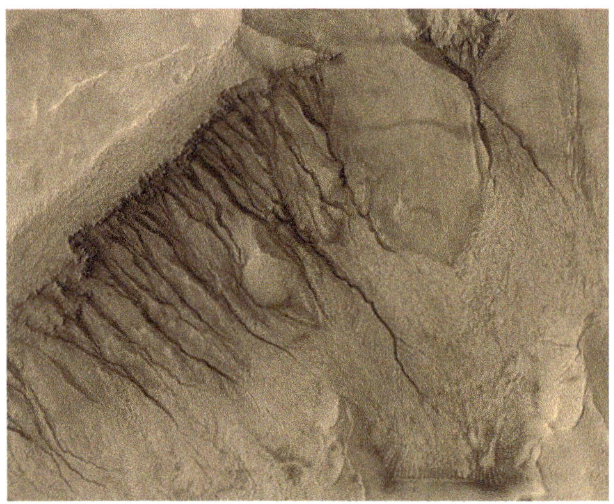

erosional features on the surface of Mars (NASA photo)

Recently, outer space has provided us with yet another reason to allow discussions of subterranean water. When the Cassini space probe flew past Saturn, it gathered data about several of Saturn's moons, including Enceladus. Cassini discovered water-rich plumes jetting from Enceladus's south pole and was able to fly through one of them to gather data. Using its cosmic dust analyzer, Cassini was able to determine that the plumes contain not only H_2O but also hydrogen gas, H_2, and sodium chloride, NaCl (salt). The mechanism driving these cold "geysers" is believed to be ***tidal pumping***. Enceladus experiences gravitational pulling forces from Saturn and its other moons. The back-and-forth pulling as all these bodies orbit and rotate might cause internal stresses inside Enceladus, leading to the heating of the water deep under its icy crust. Occasionally cracks might open in the crust, allowing pressurized water to spew out. As with Mars, not all scientists agree with this hypothesis; some believe the jets are caused by electrical forces. The point for us is that talking about tidal pumping and water jets is now mainstream.

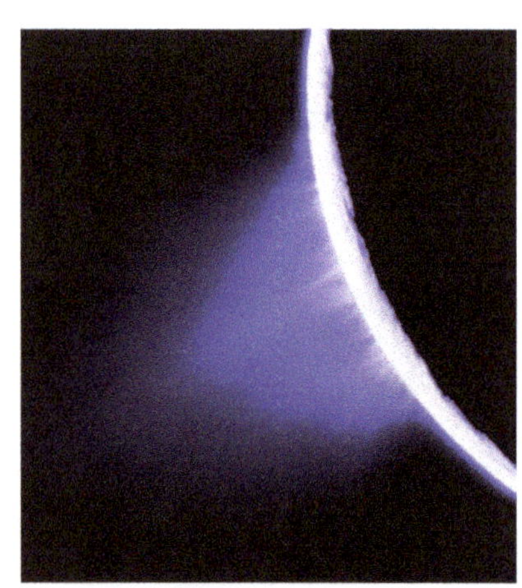

This illustration shows the jets at the southern pole of Enceladus.

Two terrestrial examples of deep water should also be mentioned, although this water may or may not be directly related to sub-crustal water. The first involves the Tibetan plateau, which lies to the north of the Himalayan mountains. In a seismic study done in 1999, researchers recorded the behavior of seismic waves (earthquake shock waves) passing through rock underneath the plateau. After analyzing the patterns made by the seismic waves, the researchers came to the conclusion that there was probably a layer of salty water lurking 15-40 km below the surface. In another study published in the journal *Science* in 2001, researchers used electromagnetic waves to probe the depths beneath the plateau. They came to basically the same conclusion, though they added that it could be a combination of water plus melted rock. The results of these two studies were surprising. Geologists and geophysicists had always assumed that it was impossible for water to exist at these depths.

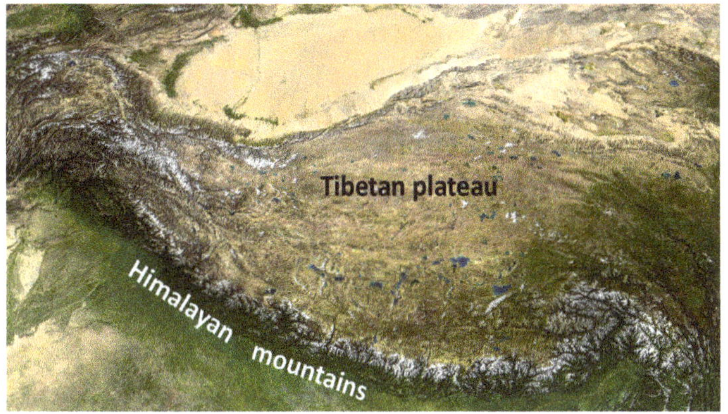

The other example of water at unexpected depths is the deepest hole ever drilled: the Kola Superdeep Borehole. Five holes were drilled (from 1970 to 1989) by Russian scientists in the Kola Peninsula, just east of Norway and Finland. The holes were only 23 cm (9 in.) in diameter but the deepest one reached 12 km (7 miles) deep. What they found at the bottom of the hole was very surprising. The drill brought up hot, salty water mixed with crushed granite. No one had predicted water at this depth.

the tower at Kola that contained the drilling rig

After this, salty water was also found at the bottom of the world's second deepest borehole: the KTB in Germany (1987-1995). The water got saltier the deeper they drilled. The water also contained substantial amounts of the elements calcium and strontium, as well as dissolved gases such as methane and carbon dioxide. The scientists said they were very sure that this deep water had not leaked down from the surface. "The fluid probably came from a deeper level and was "pumped" by tectonic forces to its present position."

Deep saltwater is also found under the state of West Virginia. People who live on the land above these areas can pump out the salty water and use evaporation to extract the salt which fetches high prices in specialty markets.

These brief adventures into outer space and down into boreholes allow us to see that it is not unreasonable to suggest that some of earth's surface water was originally located under the crust. Let's allow this possibility and see where it leads. Our rabbit trail continues...

One scientist has spent more time thinking about the idea of "primordial" subterranean water than anyone else ever has. Dr. Walter Brown, PhD (mechanical engineering, MIT) began developing a new theory in the early 1980s. He postulated that the "great deep," from Genesis 7:11, was a 1-mile (1-2 km) deep layer of water under the crust, at a depth of perhaps 10 miles (16 km). (Later, he increased his estimate to 60 miles (95 km).) This became one of the very few initial conditions of his theory. Since geologists and astrophysicists struggle to explain where earth's first water came from, even resorting to the unreasonable proposal that comets brought all if it, Dr. Brown decided to assume that an intelligent Creator had designed the earth, and part of that original design was a layer of water deep under a granite crust. The crust was like an unbroken egg shell, so the idea of a super-continent (e.g. Pangea) was discarded. There might have been very large seas, but they would have been on top of the crust.

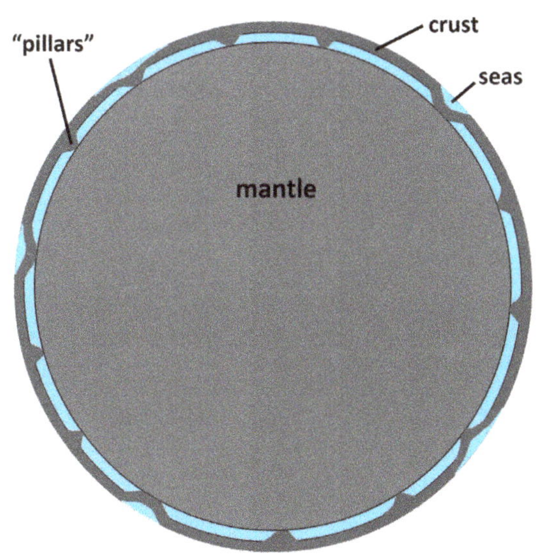

the initial conditions of Dr. Brown's theory

The pre-flood world likely had mountains, though probably not as tall as today's tallest mountains. We can't know what type of rock was close to the surface, but we can be sure that there was plenty of soil on top of it to support a large variety of plant life. There might also have been mud on the seafloors and sand on the beaches. If there was any limestone, it would have been inorganic and non-fossilerous.

Another important aspect of Dr. Brown's initial conditions is the assumption that the center of the earth was not as hot as it is today. All other theories assume that the center of the earth has always been hot, and most theories suggest that the entire globe used to be made of molten (liquid) rock. We are used to seeing artists' illustrations of the early earth looking like a sea of red lava with a few volcanoes dotted about. The molten earth idea comes from the general theory of star and planet formation, called the Nebular Hypothesis. It is assumed that after the Big Bang created atoms, gravity caused some atoms to clump together. Many collisions would have occurred as this clumping happened, and all these collisions would have generated a lot of heat. Thus, as the earth gradually formed, much heat was present. (There are many problems with this theory, but we're going to set that aside for now and not add another rabbit trail.)

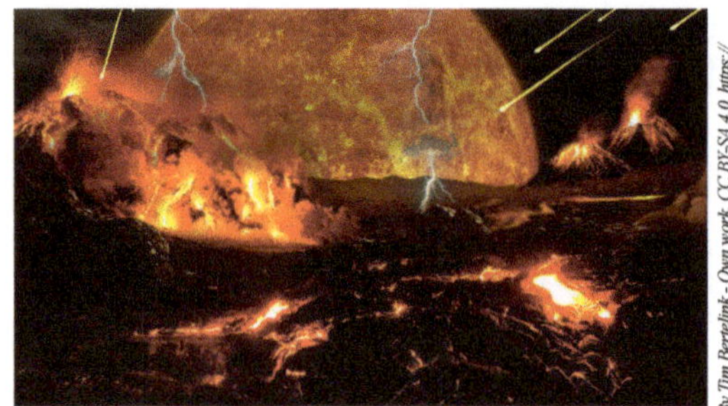

This is an artist's conception of the molten earth idea. The molten moon looms large in the background.

Behind all molten earth ideas is the fundamental assumption that the earth is one big "accident" in space, having been created by chance, not by an intelligent mind. If the earth is a creation, not an accident, then we are free to consider other options. If we consider the fact that the Bible describes the newly created Earth as "very good," we should ask ourselves whether it is really necessary to assume that the inner earth has always been extremely hot. Could mild geothermal heat have been produced another way? Volcanoes and lava flows are very dangerous to humans, so they would not qualify as "very good." Yes, we are fascinated by them and they can produce very interesting landscapes, but people who've had their houses destroyed by them can agree that they miss the mark as a "very good" phenomenon.

What would happen if a geological theory started with a much cooler earth—an earth without a liquid outer core, and without magma chambers below the crust? Could such an earth still end up in the condition it is in today? This is the challenge that Dr. Brown's theory faced.

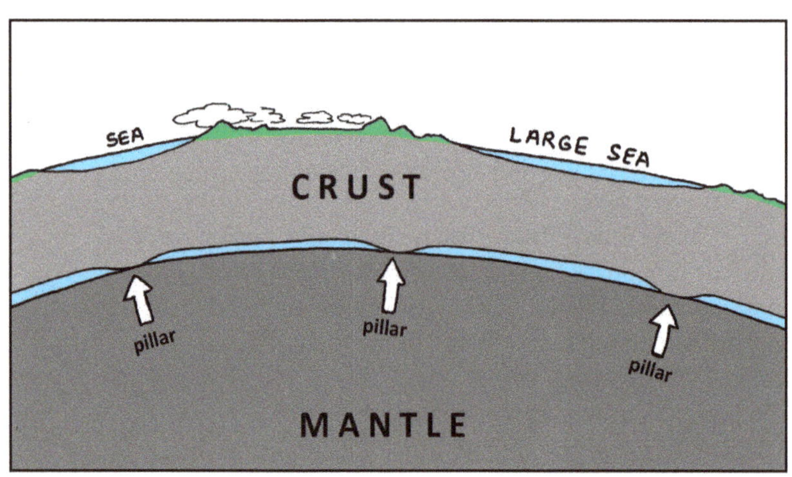

There are two more aspects of the initial conditions that need further explanation. First, something had to stop the crustal shell from sliding around on top of that slippery layer of water underneath. There must have been large supporting pillars of rock loosely connecting the crust to the mantle. The pillars probably just rested on the mantle without being actually attached to it, allowing the crust to gently flex up and down as it experienced the gravitational pull of the moon. Second, it is necessary to assume that the water under the crust would have been salty, like the water found at the bottom of the boreholes, and that it also had had a sizable amount of carbon dioxide dissolved into it. The design reason for the presence of these elements (sodium, chlorine, calcium, carbon) in the water is not entirely clear yet, but the results of the superdeep borehole research indicate that it is not unreasonable to include these elements in the sub-crustal water chamber, and the presence of these elements can provide a pathway to creating limestone and other sedimentary rocks.

Now we have the initial conditions set, and we are ready to apply known laws of physics and chemistry and see what happens. We are getting close to the end of our rabbit trail and will soon see how limestone might have been made and distributed around the globe.

The water that was trapped deep under the crust would have been under extreme pressure due to the weight of the rock pressing down on it. When a solid, liquid, or gas is squeezed, its temperature goes up. We can observe and measure this in lab experiments and develop graphs and equations that quantify just how much the temperature will go up in various situations. The weight of the crust would have elevated the pressure in the water chamber to about 372,000 pounds per square inch (25,550 bars), and raised the temperature to well above 374° C (705° F), possibly as high as 700° C (1300° F). We don't need to know exactly what the temperature was, only that it would have been over 374° C, because at this temperature something interesting happens.

At normal atmospheric pressure (about 14.7 pounds per square inch) water boils at 100° C (212° F). However, at 3,200 psi, it boils at 374° C (705° F). Above this pressure and temperature combination, water goes into a phase called "supercritical." In this phase, water molecules are so energetic that they can only form short-lived microscopic droplets consisting of a few dozen water molecules. (In steam, the water droplets are much larger than this.) No sooner than a micro-droplet is formed, it breaks apart again. Other substances can also become supercritical. In fact, supercritical carbon dioxide is used in commercial processes where a solid must be dissolved quickly and thoroughly. Supercritical substances are known for their dissolving power.

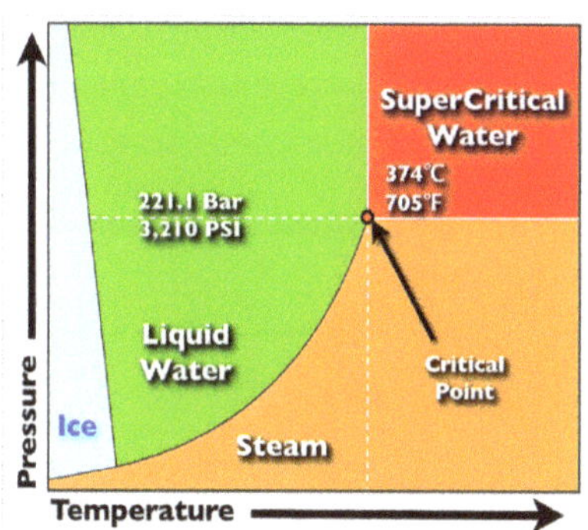

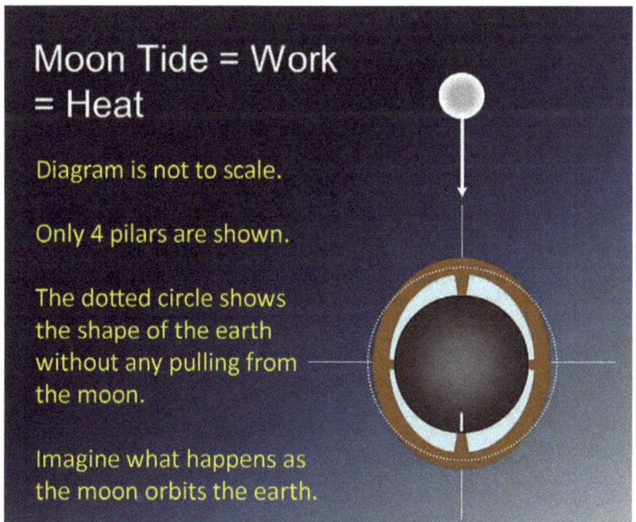

The sun also pulls on the earth helping to cause tides. The moon has twice the tidal force of the sun because it is close.

Another factor that could have contributed to the heating of the water in the subterranean chamber was the pulling effect of the sun and moon, causing ***tidal pumping***. Today, the gravitational pull of the sun and moon cause the tides we see in large bodies of water. The sun and moon pull on the entire crust, not just the oceans, but we can't see it because the solid rock of the crust is stiff and resists being pulled out of shape

In our initial conditions scenario, the crust of the earth is resting on a "water bed," and on some pillars that are resting on the mantle. Because water is squishy, and because the pillars were merely resting on the mantle, the crust would have been able to flex up and down a bit as the sun and moon pulled on it. This flexing would not have been noticeable to anyone living on the earth at that time; the ground would have seemed solid and unmoving. Even an observer high above the earth would not have noticed the flexing.

In certain materials, it takes surprisingly little motion to generate a lot of heat. (Try bending a paper clip at one place multiple times and notice the heat that is produced.) Just the gentle gravitational pull of the moon would have been enough to a cause considerable heating effect in the crust. The role of the water was to absorb the heat generated by the crust. As months and years went by, the temperature and pressure would have continually risen. Critics of the theory like to point out that this sounds very much like a planetary "ticking time bomb," which would not have been very good. This, they feel, disqualifies the theory from being compatible with an intelligently designed earth. However, there was a way for this heat to be dissipated until an equilibrium was reached and the temperature stopped rising. The key to reaching equilibrium was the dissolving power of the supercritical water and the ability of the crust to transfer the heat toward the surface.

Supercritical water micro-droplets were constantly forming, falling apart, and reforming. A droplet, while it existed, was able to dissolve a tiny bit of rock—perhaps a piece made of only a few dozen atoms. Water is able to dissolve things because its molecules carry an electrical charge: positive on one side and negative on the other. These electrical charges pull on atoms in crystals that also carry an electrical charge (either positive or negative) and the pull is strong enough to separate some of the atoms from their parent crystal.

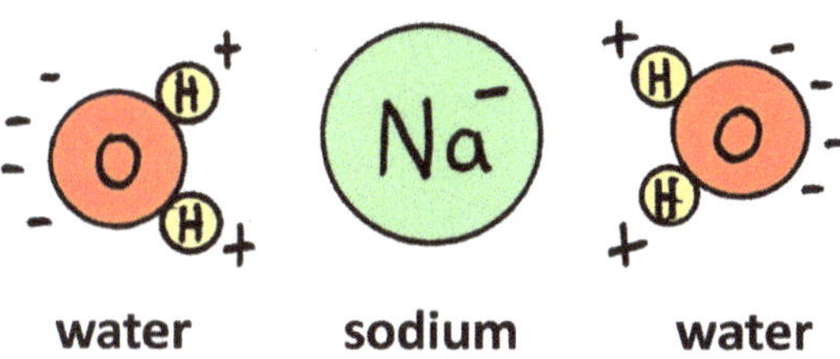

Here we see the dissolving cycle that could have been happening in the water chamber. 1) A SCW droplet forms. 2) The droplet dissolves a tiny bit of mineral. 3) As long as the droplet remains, it carries the mineral molecule. 4) The droplet falls apart due to the high (heat) energy of the water molecules. The mineral molecule is released and slowly drifts down to the floor of the chamber. We know this process does indeed happen because researchers have observed it in lab experiments. The researchers called it ***out-salting*** because it looked like white salt was falling out of the water. They also found that if they let the pressure drop, the out-salting occurred even faster. This fact will become important later on. 5) The water molecules reassemble to form a new droplet. This new droplet is free to go over and pry another tiny mineral molecule away from its parent crystal.

This cycle was repeated over and over again. Eventually, there would have been quite a pile of loose mineral slush accumulating on the chamber floor.

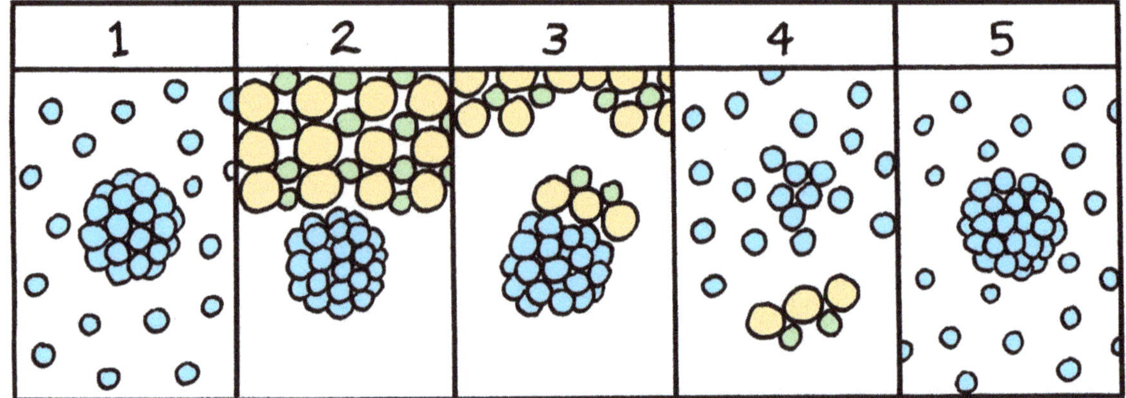

The deep crust of the earth is believed to be granitic, being made of granite or granite-like rocks. This close-up view of granite shows it is a blend of many mineral crystals.

The ceiling of the chamber would not have been perfectly homogeneous. At the molecular level, substances naturally have slight variations. Even something that has a very uniform appearance, such as a smooth steel surface or a sheet of window glass, would not look so uniform under extremely high magnification. This is especially true for a substance that is a blend of many different types of atoms and molecules. This would have been the case for the rock that formed the ceiling of the chamber. If this rock was anything like modern crustal rock, it would have been a blend of molecules made up of various elements like silicon, oxygen, magnesium, iron, aluminum, calcium, sodium, potassium, plus small ("trace") amounts of metals such as manganese, strontium, and titanium. Some areas would have had a little more of one thing and a little less of something else. The varying molecular composition of the rock would have made certain areas more or less susceptible to the dissolving power of the supercritical water. In places that were more soluble, the SCW would have eaten into them more quickly. The more the SCW ate away at susceptible minerals in the rocks, the larger these voids in the rock would have become. Over time, the amount of rock that was in direct contact with the SCW (the total surface area) might have increased exponentially as the porous voids spread within the structure.

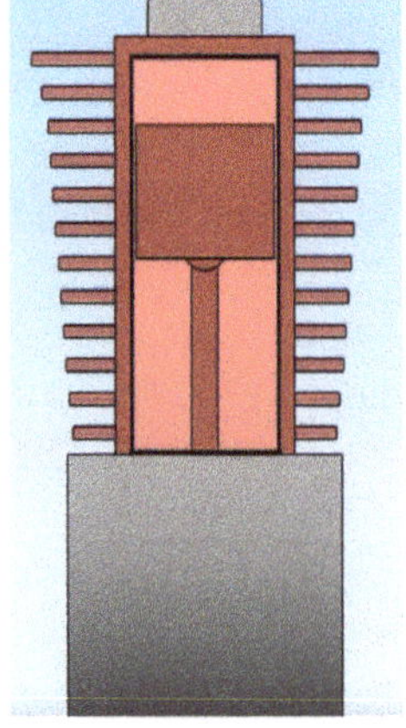

Surface area is a key to cooling. In the realm of biology, it has been observed that desert mammals often have large ears, providing increased surface area for getting rid of body heat. Our lungs are made of millions of tiny sacs called alveoli, which give the lungs a total surface area the size of tennis court. Notice how warm your breath is when you exhale. In combustion engines, "fins" can be added to the outside of a cylinder so that heat can be quickly and efficiently transferred to the surrounding air. Metallic "foams" (instead of fin shapes) are also used for cooling machines. The chamber's porous ceiling could have developed a huge amount of surface area that provided a very efficient way for the heat within the chamber to be transferred to the rock above it. The rock would eventually heat up, but the water would also be cooling at the same time.

The heat that was transferred to the rock ceiling might then have been transferred to the slightly cooler rock on top of it. Heat might have been passed along through the rock until eventually the heat was transferred all the way to the surface.

In this diagram, the blue arrow shows the gravitational attraction between the earth and the moon, and the SCW is colored red to show its intense heat. The gradation from orange to yellow shows the decreasing temperature as heat energy is conducted towards the surface. The wavy yellow lines show the final bit of heat slowly dissipating into the atmosphere, and eventually into outer space.

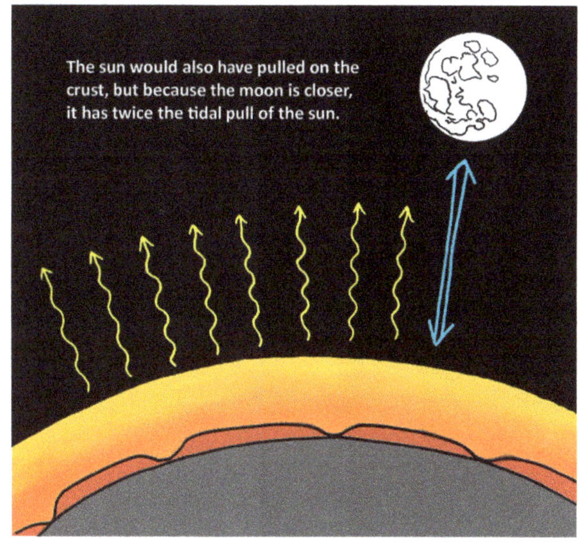

Some critics of this theory see a theological problem with assuming that supercritical water under the crust was part of the original design of the earth. They prefer to assume that the water under the crust only became supercritical after Adam sinned, and that the original plan had been for the inner earth to remain cool. There isn't a way to find out for sure what the initial conditions were. We're all making assumptions. However, if the Creator's original plan for keeping the earth warm was to harvest energy from gravity by using the rotation of the earth, the orbiting of the moon, and the flexing of the crust, then water was the ideal substance to put under the crust because it has a very high heat capacity, meaning it takes a lot of energy to increase its temperature. Either way, the end result was the same—the earth became primed for a catastrophic flooding event.

Could the earth-moon system have also been the key to the original hydrological cycle, causing daily mists to rise and fall, or driving the flow of the original rivers? This certainly is an interesting thought. It should not surprise us that the moon may have been much more than a calendar in the sky. In biology we see structures having multiple functions. Our bones not only provide a scaffolding for our muscles, but also create blood cells and store minerals. The moon wasn't a cosmic accident, as secular astronomers tell us. God made it just the right size and put it at just the right distance. As an added bonus, this combination of size and distance also makes eclipses possible. As far as we can tell, no other planet (even outside our solar system) has a moon that can make a perfect eclipse.

We've finally come to the end of our rabbit trail! We are now ready to begin making limestone. On page 6, we looked at the equation that governs the creation of calcium carbonate ($CaCO_3$):

$$Ca^{2+} + 2(HCO_3^{-1}) \leftrightarrow H_2O + CO_2 + CaCO_3$$

Calcium (Ca) was one of the elements that would have been in the water of the great deep, due to the dissolving of the granitic ceiling rock. When calcium atoms are dissolved in water they lose their two outer electrons, and turn into calcium ions with an electrical charge of +2. Bicarbonate ions (HCO_3^{-1}) are readily formed when carbon dioxide is dissolved into water, so the supply of these would have been plentiful. In this chemical reaction (going from left to right) all the atoms rearrange themselves into new molecules. Calcium's +2 charge attracts the negatively charged oxygen atoms on a bicarbonate ion (which drops its hydrogen atom) to form $CaCO_3$. The hydrogens from the bicarbonates join with an oxygen to form a water molecule, and one bicarbonate is converted into a CO_2 molecule.

If this reaction happened at the surface (in a warm, shallow sea) the CO_2 molecule would eventually end up in the air or water. Trapped in the chamber, however, the CO_2 would simply dissolve back into the water and create more bicarbonate ions. These bicarbonates could combine with calcium ions to make more calcium carbonate particles. The carbon atoms would be recycled over and over again. Because this process happened in an underground location that was sealed off from the surface, the carbon dioxide was never released into the atmosphere. The end result was not a build-up of carbon dioxide, but the formation of a tremendous amount of out-salted minerals lying on the floor of the subterranean chamber.

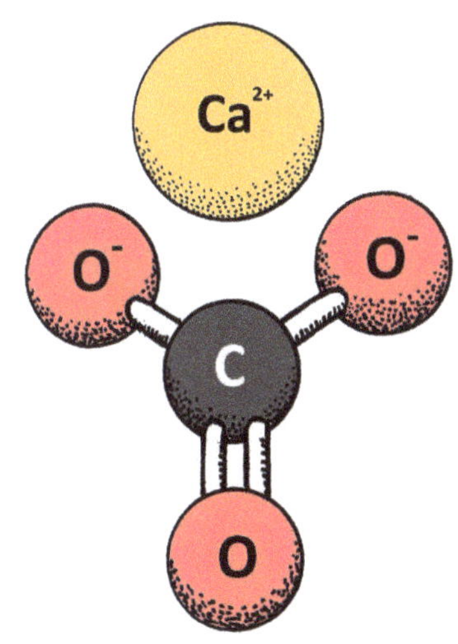

$CaCO_3$ ("limestone")

Limestone was not the only mineral that could have been out-salted onto the chamber floor. Judging by the saltiness of the brine found at the bottom of the boreholes and by the volume of salt deposits found deep in the earth, the subterranean water could easily have contained a lot of sodium chloride that might have precipitated out as tiny flakes of white salt.

There would also have been dissolved silicon dioxide, SiO_2, in the water, since granitic rocks contain a lot of SiO_2. Under normal surface conditions, silicon dioxide dissolves into water very slowly. Rain dripping over granite can only produce water with an SiO_2 concentration of 6 parts per million (ppm). Boiling water dripping over granite can up the concentration to 60-80 ppm. The only way to get concentrations higher than this is to combine high temperature with high pressure —exactly the conditions that would have existed in the chamber. Under these conditions, concentrations of SiO_2 in the water can go up to at least 140 ppm. This figure (140 ppm) is significant because experiments have indicated that this is the concentration necessary to petrify wood. If these silicon-saturated waters were released to the surface, they could easily have created the petrified logs we find in various parts of the world. Other geological theories don't have an adequate mechanism for creating super-saturated silicon solutions. There just isn't enough pressure and heat at the surface.

Because of the mineral mush blanketing the floor of the chamber, the SCW would have had less of an effect on the mantle rock below. (Additionally, the ***specific heat capacity*** of mantle rock is thought to be higher than that of granite, meaning it heats up more slowly.) Some of the mantle would have been dissolved, especially at first, but as the mush got thicker, the mantle rock would have been affected less and less.

The mantle rock that did dissolve would have released a considerable amount of silicon, oxygen, iron, magnesium, calcium and sodium into the water, and lesser amounts of nickel, aluminum, chromium and cobalt. How are we able to suggest this with any confidence? Though we've never been able to drill down into the mantle to sample it, there are clues at the earth's surface, at places where it appears that the mantle has risen up. For example, a few islands in the Atlantic ocean (such as St. Peter and St. Paul Rocks) appear to be mantle rock that was not melted into basalt. These areas show that the mantle is primarily made of a silicate mineral called ***olivine***, which is high in magnesium and iron. We call this type of rock "mafic." ("Ma" for magnesium, and "f" for ferrum, the Latin word for iron.) When olivine is mixed with a very similar mafic mineral called ***pyroxene***, it forms a rock called ***peridotite***. When peridotite softens and then cools and hardens slowly, it can form a coarse-grained rock called ***gabbro***. These Atlantic islands, as well as other parts of the seafloor, often contain much gabbro.

We can melt olivine and peridotite in a lab to produce basalt, so when we see basalt covering much of the ocean floor, we can be fairly sure that this rock started out as olivine and/or peridotite. As supporting evidence, occasionally we find chunks of basalt that contain a piece of unmelted olivine or peridotite (shown in the second picture below.) The assumption is made that this ***xenolith*** is a piece of the "parent" rock that somehow escaped melting. Gabbro, on the other hand, is thought to have been produced over long periods of time, so we can't watch it forming in a lab. Even so, because of what we know about the chemistry of mafic rocks, it seems very reasonable to assume that gabbro is made of mafic minerals that have been modified by heat or pressure, or both.

pure olivine

olivine xenolith embedded in basalt

peridotite

gabbro

If the raw materials for limestone and sandstone were being prepared in the chamber, the presence of iron in the SCW (from mafic minerals such as olivine) can provide an explanation for why iron oxide is so often found in these sedimentary rocks, staining them red. Also, the dissolved magnesium in the water provided an abundant source of magnesium ions for making dolomite. A magnesium atom will happily hold on to a carbonate (CO_3^-) molecule, just like calcium does. If magnesium was mixing with CO_3 molecules while still in the chamber, this provides a mechanism for making great quantities of homogeneous dolostone. We don't need to depend on the trickle of magnesium through already-formed limestone.

Iron gives the Redwall Limestone its color.

The volume of mineral "mush" lying on the chamber floor could have been immense. We can use math equations for volume and surface area to figure out how much there might have been and then compare this to the estimate of how much sedimentary rock there is covering the continents. The average depth of sedimentary rock is about a mile (about 1.5 km). In some places it is much deeper and in other places there isn't any at all. The subterranean chamber could have produced a large part of the sediments, but not all of them. We still need another sediment-generating mechanism, especially for producing large amounts of shale and sandstone.

As a side note, other flood theories struggle to come up with a mechanism for producing enough sediment to account for all the sedimentary rock layers. They often invoke fast erosion produced by the waves of rising flood waters. Even the rushing waters of Niagara Falls can't erode more than a few feet of rock per year. The waterfall has been able to erode several kilometers of rock over the centuries because the sediments are quickly carried away, exposing fresh rock to erode. In a global flood situation, the eroded sediments would likely drift downwards and begin blanketing the surface of the earth. As sediments piled up, the surface would no longer be exposed to erosional forces, and erosion would eventually stop. We must also ask how much erosion of rock is observed when a tidal wave or tsunami occurs today. Certainly they have great power to move loose sediments, but how much solid rock is eroded? Common sense tells us that it's not enough erosion to produce a global layer of sediments a mile deep. A better erosional mechanism is needed.

We're now ready to release the supercritical water and its load of dissolved minerals. All we need is a tiny crack in the granite shell. How might it have started? Did a weak place begin to form in one part of the shell? Some critics of this theory accuse it of setting up a scenario where a failure of the crustal shell was inevitable. God would not have created an earth doomed to disaster, they say, and therefore the entire theory is discredited. However, this critique is not valid because the initial conditions don't automatically lead to disaster. As previously discussed, the initial conditions set up a perfectly balanced system that could have theoretically continued to function forever unless some element changed. Obviously, something changed. Exactly what changed is an unsolvable mystery. A crack developed for unknown reasons.

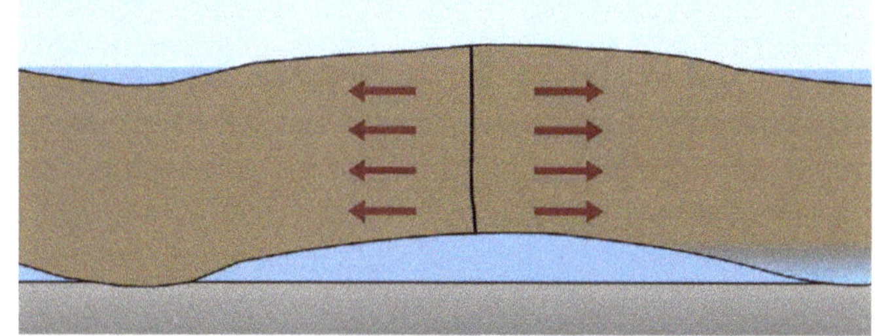

Mechanical engineers are quite familiar with how materials react to stresses and strains. They study cracks, and have learned through experiments that once a crack develops, the weakest points in most materials will be at the ends of the crack. Anyone who has observed a rip develop in a pair of pants has seen this principle at work. Once a rip starts, other rips won't start; the first rip will just continue to grow. Thus, we can confidently assume that once a crack developed in the granite shell, the crack continued to lengthen at both ends due to the immense pressure coming from the SCW below. Engineers who work with granite can calculate how fast this crack could have opened up: about 3 miles per second. The crack could have raced around the globe in only a few hours.

We can guess where this crack went by looking at today's mid-ocean ridges. (The reason for this will be discussed soon.) What the original shell looked like at the time of the crack is unknown. Places on the globe that are now covered by ocean might have once been land. Perhaps a future researcher will come up with a way to determine what the pre-flood crust looked like, but for now we'll have to leave it as an unknown. This illustration shows Dr. Brown's guess as to where the crack started—somewhere in what is now the Arctic Sea (red X). The red circle shows a place in what is now the Indian Ocean where one end of the crack eventually met up with the other end. Cracks can't jump over other cracks. You can demonstrate this with a piece of paper. Have two people hold on to perpendicular edges, perhaps the top and one side, and try to simultaneously rip the paper in half. The rip that is a split second faster will make it all the way across, but the slower rip will stop right where it approaches the place where the faster rip already crossed. Thus, the T-shaped triple junction in the Indian Ocean shows us that the northwestward-bound crack (going towards Arabia) traversed that place before the northeastward-bound crack (going toward Indonesia).

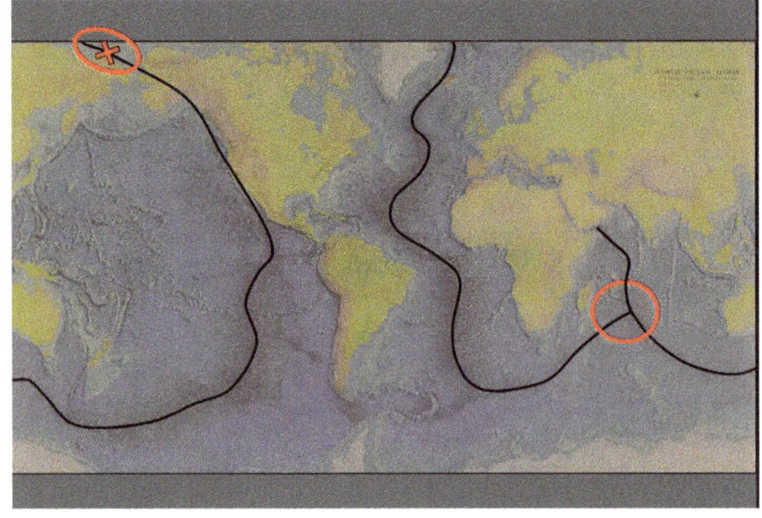

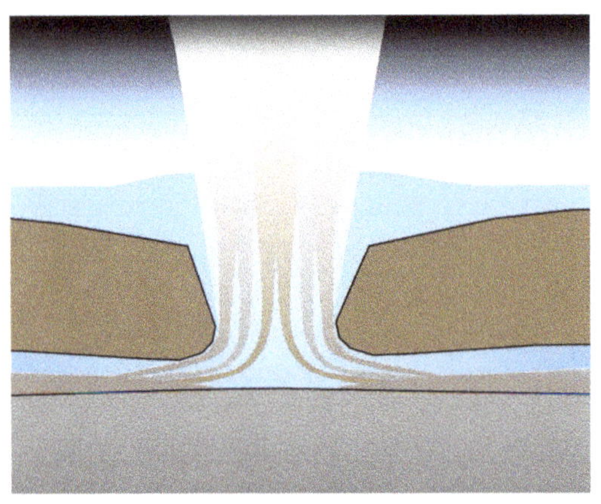

As the cracks opened up, the supercritical water would have begun to escape with a force that is almost unimaginable. Dr. Brown calculated that it could have exploded with enough force to escape earth's gravity. Much of the jetting water, and many pieces of broken rock, could have ended up in outer space, eventually forming comets and asteroids. Thus, we have an explanation for why tiny bits of limestone and other hot-water minerals are found in comets, and why asteroids are flying rocks piles held together by water ice.

The escaping water would have been hundreds of times more powerful than Niagara Falls and was able to quickly erode the sides of the crack. If you doubt the ability of water to cut through granite, search a video platform using the key words "water saw."

Not all of the water went into space, of course. Some of it fell back on the earth creating the 40 days and nights of torrential rain described in the flood story. As the crack widened and the escaping water slowed down a bit, more of the water and its load of dissolved minerals would have remained on the earth. Could our vast quantities of sand have come from this massive erosion event? Sand is somewhat of a mystery. The erosional processes we observe today are not producing much sand. We find giant sand dunes in odd places, far from the ocean. What could have created all that sand? Since much of our sand is essentially made of tiny pieces of granite (quartz, feldspar, mica), perhaps water jets cutting through granite isn't an unreasonable explanation.

The escaping supercritical water would have cooled almost instantly due to its rapid expansion. Expansion always leads to cooling. (This is how a refrigerator motor works). Because supercritical water has so much surface area (microscopic droplets), it is able to cool even more rapidly than other substances. By the time the supercritical water reached the surface, in a matter of seconds, it was probably only mildly warm. As we mentioned earlier, scientists have observed rapid out-salting in supercritical fluids if they are allowed to expand very quickly. Thus, we can assert that the load of tiny mineral particles increased as the water escaped. Microscopic bits of limestone ($CaCO_3$) were swept to the surface. This solves the mystery of how limestone could be created without releasing too much carbon dioxide into the atmosphere. Certainly some carbon dioxide might have escaped, but not the quantity that would have been created had the limestone been formed entirely at the surface. The warm, mineral-rich water provided the chemical "cement" that was necessary to harden sediments into stone. But are minerals able to sort themselves out after being all mixed together in a mineral "soup"? Apparently so, according to a research paper published by Cambridge University Press in 2019. Researchers used a mineral solution thought to be a good approximation for the ancient waters of the Mediterranean Sea. (The research was focused on how this sea's large salt formations might have formed.) They found that limestone ($CaCO_3$) separates out and crystallizes first, followed by dolomite ($Mg,CaCO_3$), then gypsum ($CaSO_4 \cdot 2H_2O$), then sodium chloride (NaCl salt), and finally, other types of salts. Similar mineral molecules were able to find each other, so to speak, and create solid crystals as the solution evaporated.

Dolomite formation is no longer a mystery. It wasn't formed by water trickling down through limestone over vast eons of time. It was formed at the same time, and in the same manner, as limestone. In some areas, the escaping water had more magnesium in it, and in these areas dolomite formed instead of the usual limestone.

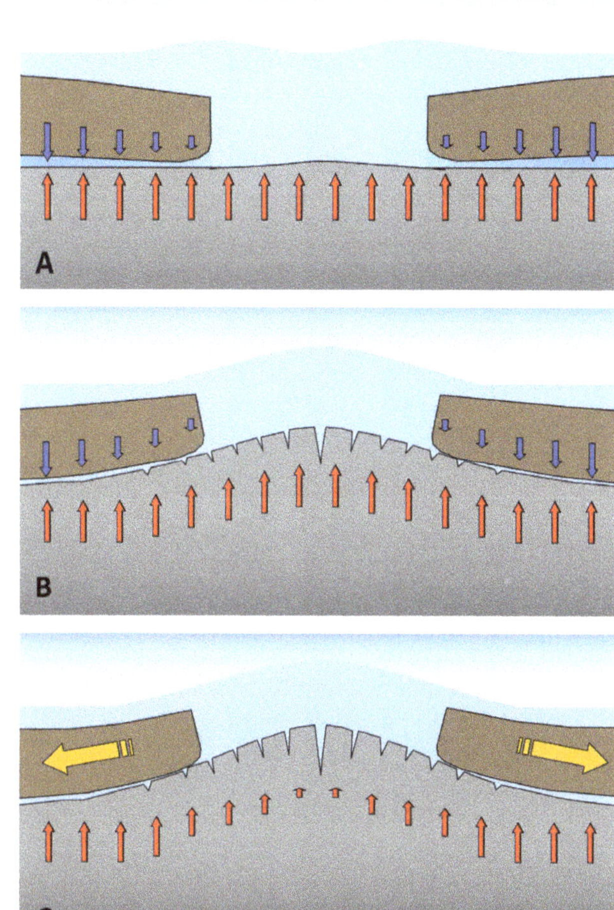

Let's look at what would have been happening to the mantle as this crack grew larger and larger due to the fast and furious erosion caused by the escaping SCW. This will let us see why we can use the mid-ocean ridges to guess the location of the crack in the crust.

As the crack widened (diagram A), the mantle, shown in gray, was suddenly free of the heavy rock that had been pushing down on it. The blue arrows represent the force of the crustal rock pressing down, and the red arrows show the counter balancing force that the mantle had been exerting against those downward forces. When there was no longer anything pressing down, those red arrow forces caused the mantle to bulge upwards, as shown in diagram B. This bulging upward is a known phenomenon called ***isostatic rebound.*** It has been observed on the floor of strip mines as rock is removed.

The upward rebound had an effect on the plates. The yellow arrows in diagram C show that the edges of the plate would have been lifted by the bulging mantle, causing the plates to slide away from the ridge. The mantle eventually rose to a height of almost 2 miles (3 km) at the center. You can see that as it bulged it also cracked. We call these cracks ***axial rifts***. However, even before these axial rifts formed, perpendicular cracks called ***fracture zones*** had formed.

In diagram D, we see that a very long line of reduced stress was forming under the arched green line. The mantle was able to bulge in this direction first, allowing the fracture zones to form. Only after the crustal crack had widened quite a bit more was there enough space for rebound to happen parallel to the crack. The central fissures are the axial rifts. The smaller cracks parallel to the axial rifts are called flank rifts.

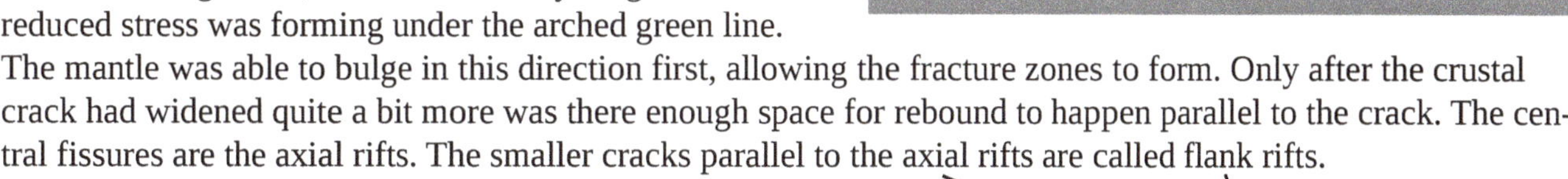

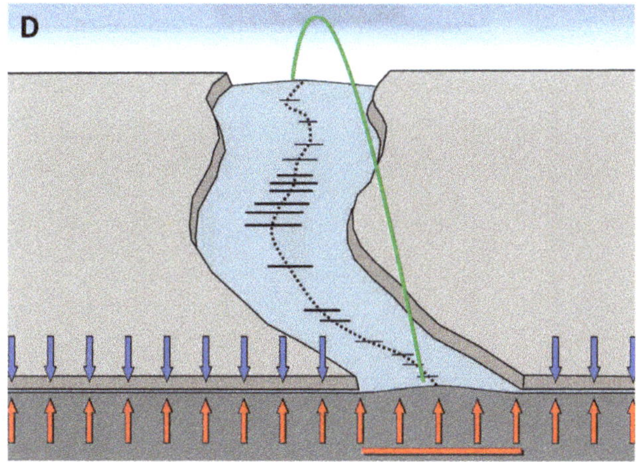

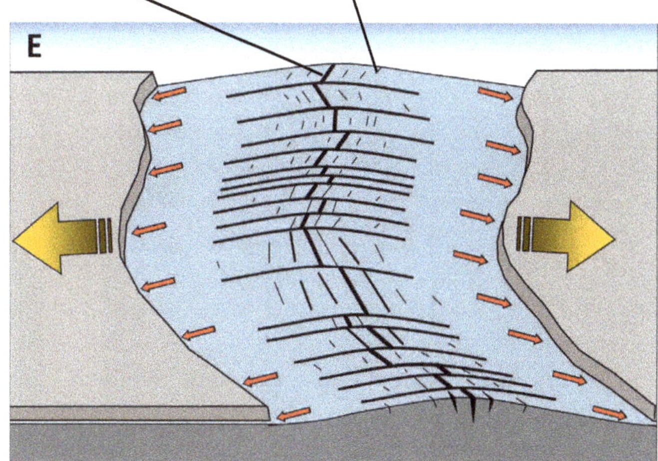

Mainstream geologists who adhere to the plate tectonic explanation of mid-ocean ridges say that the fracture lines were created by "seafloor spreading" over millions of years. They believe that the central axial rifts are places where magma has welled up from the mantle, pushing the seafloor apart on either side. They believe that the fracture zones that are perpendicular to the axial rifts are places where the seafloor cracked in order to allow for east-west movement. They imagine that these sections of seafloor have moved apart independently and at different times. The yellow arrows show the assumed movements of the sections. The tiny red arrows show the imagined spreading of the seafloor outward from the axial rifts. There are a number of problems with this scenario. The most obvious is the fact that these fracture lines end, and do not extend all the way across the ocean. Also, the ability of liquid magma to move entire ocean floors must be seriously questioned. The magma would not have been in a closed hydraulic system, so it would have been free to move in any direction. Substances under pressure always take the path of least resistance, so the magma would have spread sideways, not upwards. Additionally, to the east and west of the axial rifts we find flat seafloor. What happened to all the rifts of bygone eras? Did they just melt away? (No, the rifts are made of solid rock.) Were they covered by sediments? (Not that we can detect.) Do we see magma welling up at axial rifts today? (No.) Can we actually detect spreading happening at the rifts "in real time"? (No.) Thus, we see that the explanation being given in geology textbooks is very weak. If the explanation given in this book seems hard to believe, bear in mind how unbelievable the mainstream explanation is.

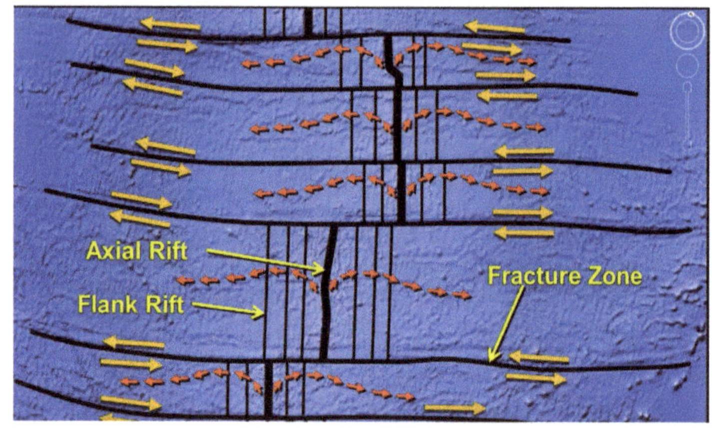

It is very hard to move a continent, much harder than geophysicists are willing to admit, though most do realize the problem that rock resting on rock presents, and that invoking magma as a lubricant, and upwellings of magma from the mantle as the driving mechanism for movement, is not a strong argument. However, they don't see any alternative because the hint they need about the mechanism that allowed motion—the layer of subterranean water—is now almost entirely gone. Seismologists detect what might be pooled salt water under the Himalayas, and we see boiling water coming out of vents on the ocean floor, but these are not strong enough clues to make us think that there used to be water under the entire crust. We see continents that appear to have moved, but we no longer see the almost frictionless watery surface upon which they used to be resting. The missing clue can only be found in a place that most scientist despise: the Hebrew scriptures. Has the bias against taking the Bible seriously prevented scientists from discovering the true geological history of the earth—the very thing they long to discover?

Because Dr. Brown's theory proposed that the continental plates (becoming separate plates after the crack raced around the globe) were resting on a layer of water, he decided to call his theory the ***Hydroplate Theory.*** The theory goes on to suggest how the rising future-Mid-Atlantic Ridge caused the continents to slide away from the ridge. It also speculates how the rising ridge affected the core of the earth, as well as what is now the Pacific side of the globe. "Nature abhors a vacuum," so the rock in the center of the earth would have shifted upwards as well, melting the inner earth and fracturing the mantle on the far side. But let's keep our focus on the rising floodwaters and how limestone and other sedimentary rocks might have formed.

As previously discussed, the escaping supercritical water cooled very quickly, according to known principles of chemistry and physics. As the water surged out from below, it would have carried with it most of the mineral mush that had accumulated over the centuries. This diagram shows that the water and its sediments continued to well up from below even after the rain stopped. The Genesis narrative tells us that the waters "continued to prevail" for 150 days, which is 110 days longer than the 40 days of rain. The fact that the water level did

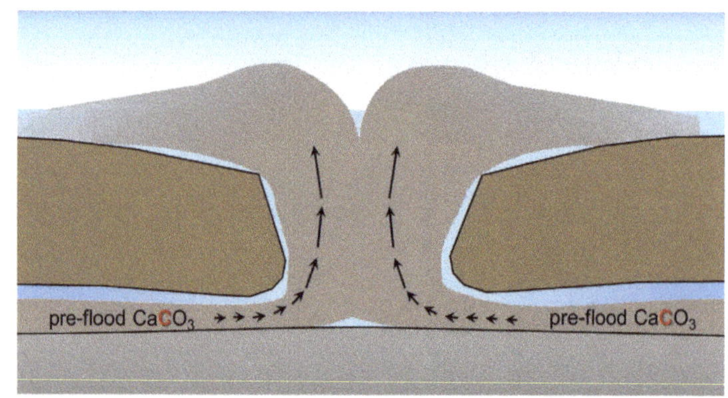

not decrease after it stopped raining, and may have even increased, is surprising unless you interpret this passage in light of the Hydroplate Theory. As the crack widened and the water pressure in the chamber started to drop, there was no longer enough force to put the water into the atmosphere. Like a hose running at the bottom of a wading pool, the water level was rising without any indication of this at the surface.

The flood waters would have been a muddy mess. Not only did they contain the mineral slush from the chamber, and from new out-salting that occurred as the water escaped due to the pressure drop, but they also had the pulverized minerals from the sides of the crustal crack, plus all the dirt, vegetation, mud and sand that had been on the surface before the flood. Additionally, there were billions of dead animals, from both land and sea. The first creatures to perish might have been the animals living on the bottom of the seas. The load of sediments could have been thick and dense enough to stay under the water instead of mixing with it, and might have rolled along the bottom, instantly covering everything with a deadly blanket of muck that would eventually harden into stone. This could have been the fate of the extinct trilobites. They appear to have been bottom-dwellers who crawled along the seafloor. The trilobite fossils we find in very fine-grained limestone are spectacular, with fine details preserved almost down to the microscopic level. Interestingly enough, very few fossils of any kind are found in sandstone. The bulk of the fossil record occurs in limestone and shale. Shale is essentially clay that has turned into stone, though it often contains more carbon compounds that modern clay does. The carbon compounds give dark gray ("black") shale its color.

a large trilobite

Now that we have sediment-laden water all over the earth, how did the sediments get sorted out into neat layers, like the ones we see in the Grand Canyon? Remember, mainstream geology claims that each layer was laid down one at a time, over millions of years, but with no explanation of where the sediments came from. Hydroplate Theory proposes that a mechanism called ***liquefaction*** contributed to the sorting of the sediments. Liquefaction suddenly came to the attention of engineers and geologists in 1964, after an earthquake in Niigata, Japan. Buildings tipped over or sank due to the liquefaction that was occurring under their foundations. Buried concrete tanks came to the surface. Liquefaction was not completely unknown before this event, but it had perhaps been under-appreciated as an important phenomenon, and one that needed to be taken into consideration when building in earthquake-prone areas. The shaking of the ground loosened all the particles of dirt, sand, and rock under the buildings, making the formerly firm surface take on almost liquid properties. The shaking also released residual underground water and allowed it to rise to the surface and come pouring out from under sidewalks and roads.

a photo from the 1964 quake in Japan

Liquefaction can be seen on a small scale when you go to the beach. If you stand on the sand where the waves are rolling in and out, you can feel your feet sink down as a wave goes back out to sea. The sand on which you are standing is full of water and as the wave goes out, the water starts to flow upwards, out of the sand. The sand grains are experiencing buoyancy due to the rising water around them. As the sand grains are temporarily suspended in water, they can shift their position. As they shift, your foot begins to sink.

Quicksand works this way, too. In fact, you can create your own little patch of quicksand if you patiently pound of a patch of wet sand on a beach. (You can search for videos of people doing this, using key words "quicksand at beach.") Vibration—your pounding feet in this case—shifts the sand grains around, allowing trapped water to escape. Even in places that don't look wet, vibration can cause hidden water to appear.

During liquefaction, particles are temporarily suspended in water and are able to shift their position. As you know, an object's density will affect how it behaves in water. If objects are less dense than water, they will float. If they are more dense, they will sink. Thus, particles would have been able to rise or sink during the time they were suspended. Size and shape can also affect how objects are sorted by water. Objects that are similar in size and/or shape will tend to collect together.

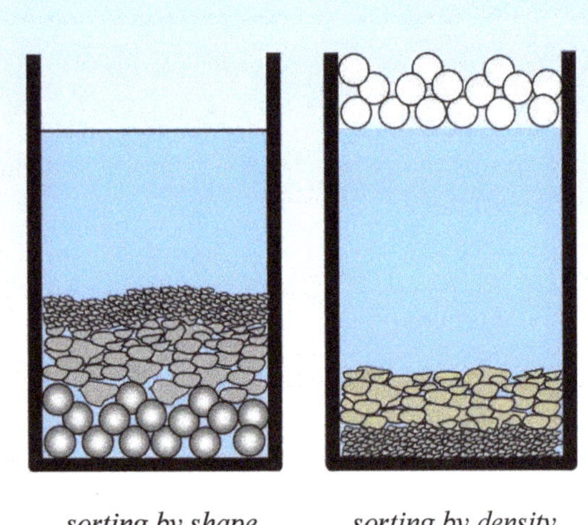

sorting by shape *sorting by density*

Flowing water gathered these pine needles from all over the driveway, then lined them so they are approximately parallel.

Quite a bit of sorting took place immediately, as the sediment particles were suspended in the water. Particles of similar density and size began to migrate to the same level. But even after they settled to the bottom, there was a way for liquefaction to still play a major role in further sorting.

The muddy, sediment-laden flood waters would have been in constant motion, not only from the force of the continuous surges coming from under the crust, but also from the gravitational pull of the moon. There would have been tidal forces generating huge swells similar to those in the open ocean but on a much more massive scale. The crests of these swells, and the depths of the troughs between them, could have reached a height of about 200 feet (about 60 meters). As a wave passed over a layer of sediments that had started to settle, the water pressure on the sediments would increase a bit, causing water to be forced down into the sediments.

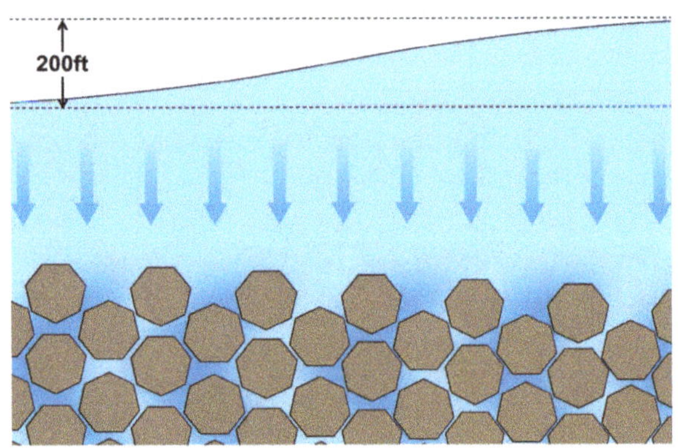

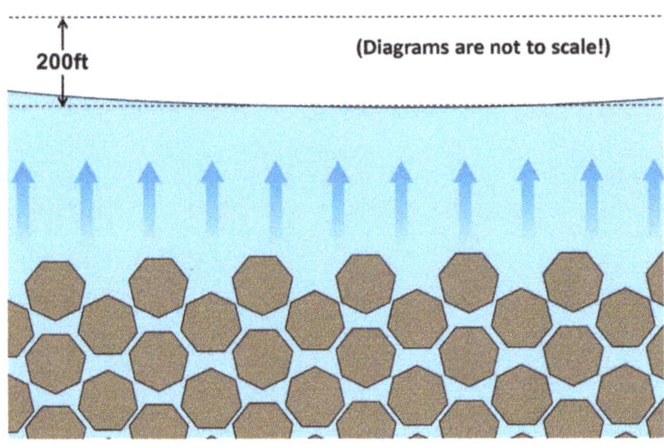

The constant back-and-forth between higher and lower water pressure over a layer of sediments would have resulted in the particles experiencing continuous lifting, then falling. Each time the particles went through a cycle of rising and falling, they would have been able to adjust their position a little more. Less dense particles could even have risen up through a layer of more dense particles, reversing their relative positions.

Once the particles settled for the last time, the dissolved minerals began to act like grout between tiles, cementing the particles together to form hard rock. The role that the super-saturated mineral solution played in hardening the rock can't be underestimated. We don't see huge amounts of rock forming today because this super-saturation was a one-time event. Contrary to what geology texts tell us, the present is <u>not</u> the key to the past.

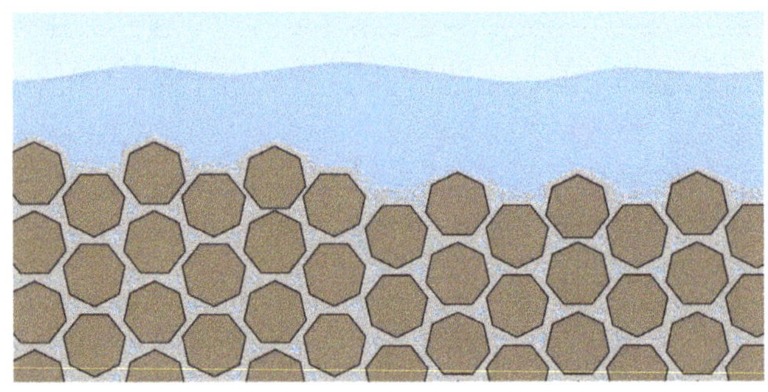

Liquefaction can be witnessed on a small scale using nothing more than two plastic bottles, a connecting hose, water, and a mixture of different types of dirt and sand. You fill one bottle halfway with the dirt mixture, and the other bottle halfway with water, connect them with the hose, then turn both bottles upside down. You then lift the water bottle higher than the dirt bottle, causing the water to flow down and out of this bottle and up through the bottom of the dirt mixture. Once most of the water has been transferred to the dirt bottle, lift the dirt bottle higher than the water bottle until the water drains back into its original bottle. Then repeat this process several dozen times. You will start to see distinct layers form as similar particles work their way into the same horizontal position. The sample bottle shown here started out with no layers at all and developed these layers quite quickly.

The sediments started out as a uniform mix.

Why have geologists overlooked liquefaction as a possible explanation for at least some of the rock layers? 19th century geologist Charles Lyell is largely to blame. It was he who made it his mission to "free science from Moses" (i.e. from the flood story in the Bible). He convinced most geologists of his day that geological processes have always occurred at the same rate we observe today.

Another force could also have been in play as the sediments were settling. The water escaping from under the plates undoubtedly came out in pulses, perhaps not at first, but as time went on there would have been more and more irregularities occurring, changing the rate of flow in various places. Water pulsing out from under a plate would have created ***flutter*** in the crust. Flutter is what flags do as they ripple and flap in the wind. A continent is stiffer than a flag, of course, so its flutter would have happened much more slowly.

The three images below show the same place at different times. Try to imagine seeing the images in succession, like an animation. The striped sediments would be undulating up and down. This fluttering and undulating would have contributed to the sorting and liquefaction happening in these water-saturated layers.

A supporting bit of evidence for flutter having occurred is the discovery of ***lineaments***. Lineaments are very straight lines that were first seen in side-scanning radar images taken from airplanes flying several miles above the ground. More recently, Landsat images have been used. The lines are believed to be radar reflections of fractures in the rock underneath all surface features (under dirt, for example). Computers are used to help analyze the data and "connect the dots" between broken line segments that are actually part of one long lineament. Sometimes the lines correspond to features that can be seen on the ground, such as fault lines, where rock has broken and one side has slidden down. Often, though, the lineaments can't be seen at all from the ground. You can walk right over one and never know it.

Everyone agrees that lineaments represent places where rock was fractured by tensile (pulling apart) forces, but the mechanism by which this happened is a mystery, especially because, in most cases, the cracks are now tightly closed, seeming to be nothing more than a scar from a past catastrophe. It is one thing to say that "uplift" caused these fractures, but it is quite another to be able suggest the mechanism for how the uplift happened. Hydroplate Theory can suggest a mechanism. As the crust was stretched upwards during a flutter, long, thin cracks opened up. When the crust went back down again, the cracks closed. According to this hypothesis, we would expect to see many parallel lineaments. As the direction of the flutter slowly changed, new series of parallel lines would be created. This matches perfectly with the data. Lineaments come in groups of parallel lines.

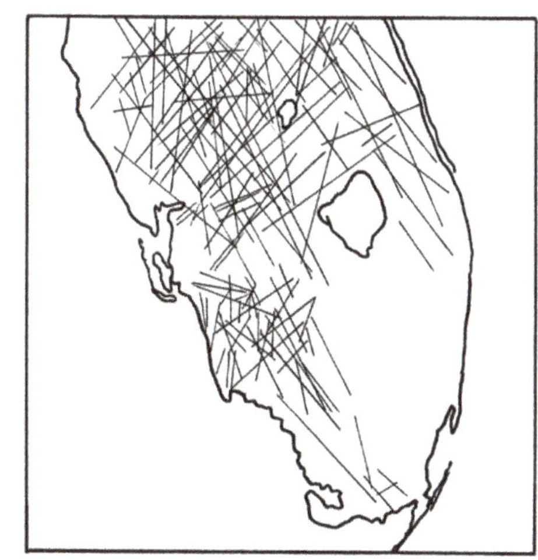

a facsimile of lineament data for southern Florida

Rock layers containing limestone often come in repeating patterns called ***cyclothems***. A typical order for a cyclothem is (from bottom to top): sandstone, shale, limestone, dense clay, fine clay. This order is consistent with the liquefaction explanation. What is surprising, however, is that this order is found worldwide, a fact that is harder to explain using theories based on local conditions—deserts and seas coming and going in a certain area. The line between the layers is often very sharp. Did a desert turn into an ocean almost overnight? If the standard theory is correct, then the rocks layers shown in this photograph would be interpreted as this area going from desert to bog to ocean, over and over again, with only one type of sediment accumulating during each eon.

black lines are coal seams

Even stranger is the fact that ***coal seams*** are sometimes found between cyclothems. As many as 50 coal seams have been found stacked on top of each other with cyclothems between them. Coal is without question the remains of ancient plants. Some coal still has recognizable chunks of wood in it, and many plant fossils have been found in coal. Textbooks tell us that coal formed as plants in bogs fell over and died over millions of years. Gradually they were buried and then the pressure of the overlying rock turned them into coal. That would mean that each layer of coal represents a wet, boggy era before or after eras of desert (sandstone) or ocean (limestone).

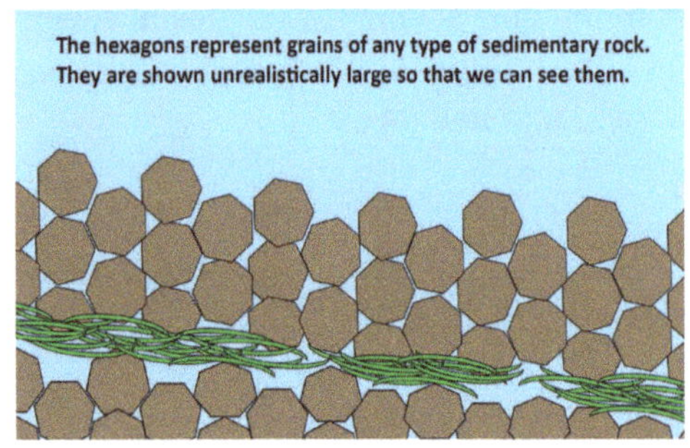

The Hydroplate Theory proposes that liquefaction helped to create both cyclothems and coal seams. The destructive powers of the fountains and the flood waters would have ripped up all of the pre-flood vegetation. (Judging by what we find in coal seams, the vegetation was lush and the plants were triple the size they are today.) Liquefaction would have sorted plants into layers at the same time it was sorting minerals. Also, fresh sediments may have flowed in from time to time, creating alternating layers of plants/sediments. Liquefaction would have been able to sort each cyclothem simultaneously.

Plants that were floating on the surface of the flood waters would have ended up as piles of soggy, rotting vegetation that, if not compressed, turned into topsoil. Some of the coal that we find near the surface might have experienced pressure from the sides, not the top. The Hydroplate Theory proposes a Compression Event after the continents slid away from the mid-ocean ridges, due to their sudden deceleration as their water cushion disappeared and they ran into friction with the mantle rock. Sideways compression can explain much metamorphic rock that is close to the surface. For example, when limestone is squeezed and heated, it turns into marble. If marble had to have formed by pressure from above, then we must require that marble quarries, like the one shown here, were once deep underground and somehow were lifted while all the rock on top of them was eroded. However, if sideways pressure is applied, there is no need to suggest that those quarries were ever deeper than they are now.

limestone coal ball

If coal seams formed in this way, instead of in bogs over millions of years, then it is not surprising that someone found a shark jaw in a coal mine in Kentucky. The flood waters would have had pieces of dead animals floating in them. It is also not surprising that a coal mine in Illinois found what looks like a fossilized log mat made of the stems of gigantic extinct plants similar to our modern day club mosses and horsetails. Nor is it surprising to find "coal balls," which are blobs of rock mixed into the coal, often made of limestone that is stained red with iron. Many coal balls contain fossils, most often of leaves or bugs, but infrequently of small sea creatures. The presence of sea fossils in the balls and the insistence of some geologists that interaction with sea water had to have played an important role in the formation of the balls, has led to quite a debate about whether the balls formed "in situ" (right there where they are found) or whether they formed in a marine environment and then were washed by storm water into a coal forming area. The in situ advocates point to examples of multiple coal balls along a single tree branch, and fragile bits of leaf sticking out of some coal balls, both of which are unlikely to have survived a rough and tumble ride in storm water. Hydroplate Theory favors the in situ explanation but explains why limestone, sea water, and marine life are found in close association with coal. It also explains why iron, sulfur, and magnesium are found in these calcite balls. Both the coal and the limestone were being formed at the same time in the midst of mineral-rich water. The calcite coal balls came from nearby layers of calcium carbonate. All these things solidified in the years following the flood.

a calcite coal ball from Illinois

This is more speculative, but animal life could have gotten trapped in pockets of water called ***water lenses*** that developed between the layers during liquefaction. The pocket might have been due to a layer of matted vegetation that was much less permeable to water than the sediments were. (The plants would have formed the "roof" of the water lens.) Or, as seen on a very small scale in bottle experiments, water pockets might have developed between sediment layers even without vegetation being present. We can't know for sure if this happened, but the effect of water lensing might help to explain the mystery of why we find fossilized tracks of reptiles and amphibians lower in the fossil record than their fossilized bodies. This observation is a problem for old-earth theories. How could millions of years have passed between the time the tracks were laid down and the time that the bodies were preserved? If water lenses occurred during the flood, perhaps small animals trapped in the lenses were able to walk around on the sediments beneath their feet for a short time before they were swept upwards as the water in the lens flowed to areas of lower pressure located above them.

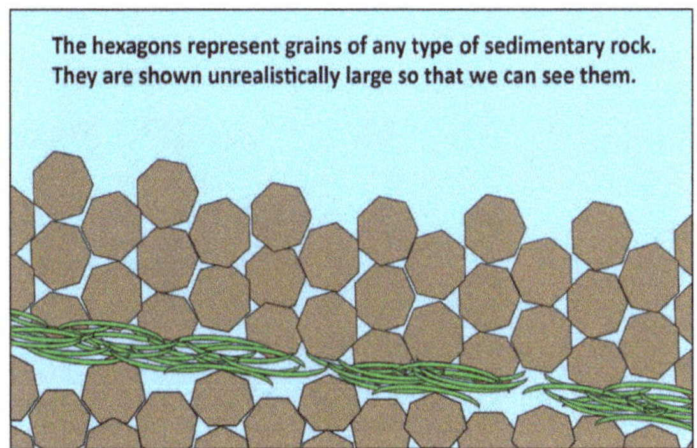

proposed water lensing

The water lens idea could also help to explain why we find fossilized fish "graveyards" sandwiched between sedimentary layers. Why did millions of fish die all at once and get perfectly preserved in rock? When fish die naturally, they float, and much of the time are eaten by scavengers. If their remains do sink to the bottom, they will be decomposed by bacteria in short order, and the bones will be scattered. If the present is the key to the past, why don't we see dead fish on lake bottoms that are in the process of being fossilized? However, if schools of fish got sorted into water lenses during liquefaction, the entire school would have perished together very quickly, and in water that had the mineral saturation level required for fossilization. As the water lenses collapsed at the end of the flood, the fish were flattened under the weight of all the sediments above. Thus, we find layers of limestone containing huge schools of exquisitely preserved paper-thin fish. (Even if lensing was not involved, this type of fossilization still required very unique conditions.)

How did so many fish get fossilized all at once, and with no signs of natural decay?

Why are layers of coal so flat if they are the remains of bogs? Also, bogs are shallow. Some coal seams are very deep.

Other factors may also have been in play during the tumultuous year of the flood. We can't possibly know exactly how the water and sediments came out from beneath the crust, and we don't know how much rock, sand, and dirt were on the surface of the earth before the flood. What was the original surface rock like, and did any of it get reworked into the sedimentary layers? Was there any inorganic limestone on the pre-flood surface? If so, eroding it to make new limestone would have been "carbon dioxide neutral." It's also possible that the watery sediments came out in pulses for quite some time, intermittently sending out "blankets" of various types of mush, creating some of the bottom sedimentary layers even without any liquefaction. What we can be fairly sure of is that however they formed, most of the sedimentary layers were in place by the end of the flood, because we have strong evidence that flood run-off eroded some of these new layers in various places, leaving telltale formations that show what used to be there. The amount of erosion recorded in the geological record defies uniformitarian explanations.

Trace fossils can sometimes add additional mysteries. Trace fossils are things like footprints, leaf impressions, burrows, etc., where the original plant or animal is not present. For example, we find dinosaur footprints on the ceilings of a few coal mines. (The best ones are found in Utah.) This would indicate that a great volume of plant material had already accumulated before the flood, and had formed a surface that would hold a foot impression. The footprints had to have been quickly covered with sediments while they were relatively fresh in order to have them preserved. Secular and old-earth theorists guess that a small-scale flooding event brought sediments across the top of the bog to preserve the footprints, then millions of years of sediments piled up on top of that. Critics of the liquefaction hypothesis point out that intense liquefaction cycles occurring after the footprints were made should have obliterated the prints. However, if this coal seam and the 3,000 feet of shale underneath it started out as a deep swamp sitting on top of 3,000 feet of clay (in the original creation), then we would not expect these bottom layers to have experienced much liquefaction at all. If this hypothesis is correct, we would not expect to find any significant fossils in the shale under the coal seam—only fossils of tiny animals that might have lived in that mud.

During its fluttering, the crust may have risen out of the flood waters quite frequently, especially before the water reached maximum depth. It is hard to imagine 150 days of rising water. We can safely assume that for at least the first half of this time period, the higher elevations were not yet covered. Animals that were able to escape to higher ground did so. Could this be why their footprints are found lower in the fossil record than their bodies? As the crust heaved up and down, animals would find themselves briefly able to walk, perhaps for a few minutes, before being lifted up once again by the water. A fresh load of sediments might have rolled in as the crust relaxed downward, covering footprints or trackways. The picture below shows tracks made by some type of invertebrate, perhaps a slug-like animal.

After a while, there could have been quite a mix of marine and terrestrial animals. In fact, this is exactly what we find in some "fossil graveyards" such as the Solnhofen limestone quarry in Germany, and the Green River formation in Wyoming. In other places, entire ecosystems were almost instantly buried. This seems to have been the case in the Burgess Shale formation in Canada.

archaeopteryx in Solnhofen limestone

invertebrate trackways in central Wisconsin

Shale, as we've mentioned, is basically mud that has hardened into rock. The black color of shale comes from ***kerogen***, an organic substance left by decomposing algae and bacteria. In the Burgess Shale, the vast majority of the creatures are bottom-dwelling sea life such as marine worms, trilobites, sponges, comb jellies, algae, shelled brachiopods, and many species of extinct horseshoe crab-like arthropods. How does an entire underwater ecosystem get fossilized? Soft-bodied animals like comb jellies and worms can only be preserved if they are buried in mineral-saturated sediments very quickly. The Burgess Shale is located in the mountain ranges along Canada's western coast. Hydroplate Theory proposes that these mountains rose up quickly during the Compression Event at the end of the flood, after the hydroplates slid and then ran into friction and skidded to a stop. As the plates buckled and crushed, the mountain ranges of the Americas (Rocky Mountains, Andes) rose in a matter of hours or days. Apparently, this part of Canada was covered by a pre-flood sea. The mountains lifted the sea, mucky bottom and all. The waters of the sea eventually drained away after the flood, and the mud and muck turned into shale. Only the Hydroplate Theory flood model can provide a mechanism powerful enough to lift a marine ecosystem miles above sea level in a short period of time. Slowly moving plates would not have enough force. Force= (mass) x (acceleration). If acceleration is almost zero (cm per year), as plate tectonics proposes, even a very large mass won't have enough force for mountain building.

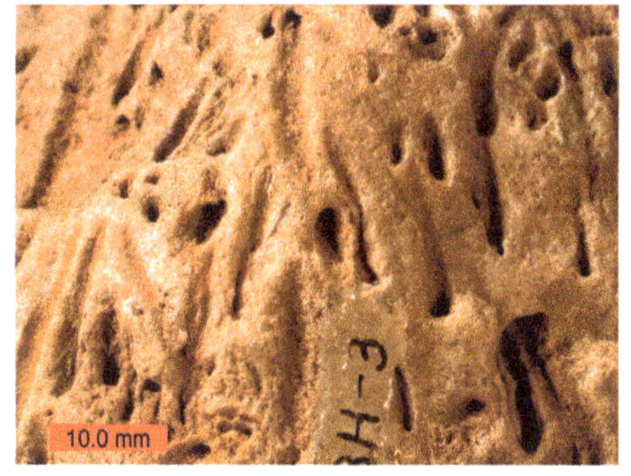

Delicate trace fossils such as leaf prints and animal burrows need to be studied on a case by case basis. Leaf prints may have survived liquefaction because during the time that liquefaction was going on, the leaf was still a leaf. Only in the years after the flood did the leaf cells rot away and leave nothing but a print. Some holes thought to be burrows could turn out to be something else, or perhaps were in a place that stayed out of the water for a longer period of time, giving animals a few days to dig before they were swept away. Fossil burrows usually occurs in sandstone, and sandstone is known to be very poor for fossilization. Obviously, in places where confirmed fossil burrows occur, liquefaction had either already occurred or did not occur, else they would have been erased. Each site would have to be evaluated according to factors such as location and type of rock. The flood was a complicated event.

MAKING COAL

Did you know that scientists can make rocks in a lab? In order to test their theories about rock formation they put raw ingredients into a small container that is able to withstand intense heat and pressure. Surprisingly, in these conditions they can often form a rock in only minutes or hours. The rock that takes the longest time to form is coal, because the reactions take place at lower temperatures. Even so, they can make coal in a few weeks. The most important lesson learned in making coal was that plants alone will not form coal. The necessary catalyst for turning plants into coal is ***clay***. Plants turn into coal when they are mixed with clay minerals then put under only moderate temperature and pressure. The flood waters provided abundant watery clay minerals that could mix with all the dead plants. Even so, giant coal seams at the surface, like the one shown here, push the limits of any theory.

All theories run into data that doesn't seem to fit their model. For every oddity that seems to be a problem for HPT, there are several oddites that seems to disprove the standard textbook theory. For example, how did this chunk of pink quartzite (circled in blue in the left photo) in one of the walls of the Grand Canyon migrate up from the pink quartzite layer below? In the photograph on the right, you can see the trail (red arrows) left by the rock as it pushed up through these layers when they were still soft. Notice how the surrounding rock looks like it formed around the block. In 1978, a geologist named Arthur Chadwick figured out what force had been able to transport

and lift this block. A thick, sand/mud/water slurry plucked the quartzite block from what was then "upstream" (to the right in the photo) and transported it "downstream." Liquefaction played a role as the sand/mud slurry experienced compression, and water and sediments were forced upward, lifting the block. The sand and mud stopped flowing, and then eventually hardened into rock, encasing the quartzite block. Despite this undeniable evidence that all the rock layers around, above, and below the quartzite block (the brown layers in the photo on the left) were soft sediments at the time this block moved into place, when these same rock layers are identified further down the canyon wall, they are said to have accumulated over millions of years. In this case, HPT provides the better answer to the question of how these rock layers formed.

In addition to liquefaction, there is another key factor to consider when attempting to explain stratified rock layers. Sediments tend to fall out in layers when they are moving in a current. Experiments have been done with a large contraption called a ***racetrack flume***. Imagine an oval, racetrack-shaped tank in which the water can be circulated around and around and around, simulating a flowing river or flood water. The tank is either clear or has several observation windows at various places so that the scientists can watch what happens in real time. They measure the rate of flow in various parts of the water; the surface water flows faster than the water at the bottom. Somewhere in the middle, a phenomenon called ***Kelvin-Helmholtz instability*** can occur. There is a fast alternation back and forth between laminar and turbulent flow. The end result of this alternation is the accumulation of many thin layers of sediments, and a layer can form in as little as one minute.

Since sediment layers can be created quickly in a flume, it is reasonable to wonder if many, or perhaps all, of the thin layers we see in the geological record might have been formed in a similar way, in flowing water, not over millions of years. Until recently, geologist have claimed that each layer represents at least a year, so we can just count the layers to determine the minimum age of the rock. Apparently this is a faulty assumption.

These layers were probably formed by alternating laminar and turbulent flows in a current of water.

What about limestone? Can layers of limestone have been created by moving water? This question was posed to Dr. Steve Austin by a colleague who believed that limestone could only be the result of a "warm, shallow sea" over millions of years. Dr. Austin believes that any type of sediment-laden water can form layers quickly. He bases his opinions on his observations of flume experiments and also on the field work he did at the Mt. St. Helens site where the eruption of a volcano produced many layered sediments in only days. Dr. Austin is also an expert in Grand Canyon geology, and is particularly interested in one specific layer in the Canyon: a 2-meter thick layer of limestone at the bottom of the Redwall Limestone that has a vast number of nautiloid fossils. A nautiloid is an extinct sea creature that looked like a squid in a long, cone-shaped shell. (Nautiloid fossils don't show the squid; only the long shell has been preserved.) This layer of limestone can be traced down through the Canyon and beyond, into other states, covering thousands of square kilometers.

nautiloid fossil

Austin estimates that the density of nautiloid fossils averages one nautiloid per cubic meter. He was the first person to notice an important detail about these fossils: 15% of them are "standing on their heads," so to speak. It looks very much like a giant school of nautiloids was caught in a slurry of limey sediments and then found themselves stuck and unable to move. If they had died naturally and slowly sank down to the seabed, the shells should all be parallel to the ground, not randomly scattered as they are. This layer of limestone can't possibly have formed over millions of years in a warm, shallow sea. Also, if the warm, shallow sea explanation is correct, why don't we find any intact structures in this layer? Any coral we find looks like it came out of a blender. We don't find large coral reefs. Those warm, shallow seas are also often described as placid. The nautiloid limestone looks anything but placid. Did density and liquefaction play a role in the sorting of the nautiloids into this layer while it was still watery? It's strange that we find so many of these animals here in this layer. Did they get sorted into this layer due to the density of their body? Or did they roam the pre-flood seas in giant schools?

Another place we find some very interesting layers of limestone is in Death Valley, California, one of the hottest and driest places on earth. Oddly enough, there is strong evidence that this valley was once a very large lake. Hydroplate Theory has no problem with Death Valley being covered with water long ago, or even being covered with a glacier for a time. There would have been many post-Flood lakes that have since dried up.

The photo on the left shows a 700-foot tall limestone butte that is sitting in the middle of a very flat valley. We'd all agree that the striped limestone layers of this rock could not have formed while the rock was in this position. The layers formed while the rock was flat, then something happened that tipped the rock into this position. The Death Valley area is in an earthquake zone and there are many visible faults where rocks have cracked and

the sides have moved in different directions. There is a major fault behind the striped butte. The rock formations behind it are made of dolomite and shale that are believed to be completely unrelated to the butte. As you can see, the layers of the butte shows huge folds, twists, and offsets. The folds show us that <u>all</u> of the layers were pliable at the same time, leading us to question the one-layer-at-a-time-over-eons explanation. How did this formation get to where it is? We've never witnessed a geological event of this scale. Perhaps the layers formed during the flood and the tipping occurred either during the compression event or in a post-flood "aftershock" event.

The photo on the right shows a limestone ***breccia*** *(bretch-ee-ah)* by the side of a road in Death Valley. Chunks of dark limestone are surrounded by white travertine (page 4 introduced travertine). The dark limestone had to have hardened, then cracked, and as the pieces slowly drifted apart, water saturated with $CaCO_3$ seeped in to fill the cracks. So this formation required two rock-forming events—one for the dark-colored rock and another for the light-colored rock. Geologists think the cracking of the dark limestone probably happened as the result of a massive earthquake. Why these patches of breccia are only in a few isolated parts of the valley is a mystery.

Now we move to the east coast of North America, to look at a very different type of limestone—a much more organic type. This satellite image shows the Bahamas islands, just off the coast of southern Florida. Land is shown with greens and browns. Cuba can be seen in the lower left corner. The dark blue color represents deeper ocean water and the lighter blue shows you where the water is very shallow, usually no deeper than 25 meters (80 feet). The shallow water is shallow because it is on top of an underwater "platform" of limestone. The thickness of the platform is estimated to be almost 3 miles (4.5 km).

As usual with any limestone formation, geologists fall back on the "warm, shallow sea" explanation for the Bahamas bank formation. Here is a quote from Wikipedia, which is usually a very good source for mainstream thought: "As the limestone was deposited in shallow water, the only way to explain this massive column is to estimate that the entire platform has subsided under its own weight at a rate of roughly 3.6 centimeters (2 inches) per 1,000 years." One has to wonder how this subsidence could have happened. How did the bedrock *underneath* the platform move down? Where did it go?

Imagine a warm, shallow sea with coral growing all over the bottom. Imagine the coral getting larger and taller, effectively making the bottom of the sea closer to the surface. If it gets too high, the coral will die because the water is too shallow. So to compensate for this increasing height, the rock beneath the sea has to go down a bit. Imagine this happening over and over again for a million years. Coral grows up, underlying rock sinks down. Against common sense, let's assume that the rock successfully sank down at the just the right rate for optimum coral growth in the sea. At the end of a million years, the warm, shallow sea wouldn't look that different on the surface, but underneath it would be hundreds or thousands of meters of limestone made of the remains of dead corals plus the shells of dead bivalves (clams) and gastropods (whelks and snails). When this scenario is applied to the Bahamas bank, how did this deep deposit of limestone manage to reverse its position and become higher than its surroundings? What pushed it up? It's easy for geologists to use word "uplift" as almost a magical incantation that can make rocks do what their theories require. In reality, "uplift" requires a massive mechanism.

Rather than create an unrealistic and complicated narrative, wouldn't it make more sense to suggest that a large load of mineral deposits (vast quantities of calcium carbonate) got dumped at this site, and then corals began to grow on top of it? Hydroplate Theory suggests that great volumes of mineral "slush" came out from under the eastern side of the American hydroplate. The Bahamas sit right at the edge of the plate, exactly where a dump might have occurred. Of course, the dump has been eroded and shaped since then, so we don't know what it looked like originally. It was a great place for shelled marine life to grow, so they flourished.

Geology texts say that after this limestone platform formed and was somehow elevated, the global sea level then fell so much that the platform stuck out of the water. During the years it was out of the water, rain and wind eroded the surface, leaving it full of small holes and caverns. When sea level rose again and most of the platform was once again covered with water, those holes became today's "blue holes." These photos show blue holes on both land and in water. The water in the land holes can be either fresh or salty, or a mix.

By Ton Engwirda - Own work, CC BY-SA 3.0 nl, https://commons.wikimedia.org/w/index.php?curid=13524568

By James St. John - Watling's Blue Hole (San Salvador Island, Bahamas) 6, CC BY 2.0, https://commons.wikimedia.org/w/index.php?curid=38371310

This is the way one website puts it: "The Bahama Banks were dry land during past ice ages, when sea level was as much as 120 meters (390 feet) lower than at present; the area of the Bahamas today thus represents only a small fraction of their prehistoric extent. When they were exposed to the atmosphere, the limestone structure was subjected to chemical weathering that created the caves and sinkholes common to karst terrain, resulting in structures like blue holes." (geoinfo.amu.edu)

The sand on the Bahamian beaches is very different from most beaches—it is based on calcium carbonate (limestone), not silicon dioxide (quartz). It is composed of broken shells, aragonite and calcite crystals, plant ash, bone, crushed coral, plankton detritus and remains of microscopic sea plants and animals.

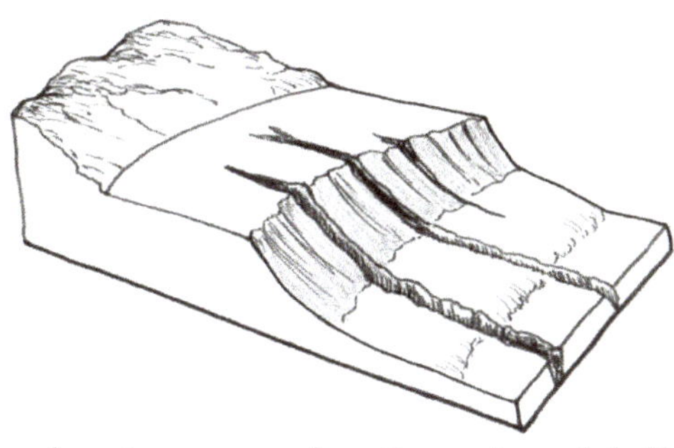

submarine canyons along the continental shelf

Most of this is very compatible with Hydroplate Theory, though HPT shortens the timeline and provides mechanisms to explain the geological processes. As the mineral mush poured out from underneath the American hydroplate, a few places got a rather large dump. As the flood waters receded (rushing into the newly formed basins, especially in the Pacific), many places that are now underwater were above water. The edges of the continental shelves, which are today under water, were the actual edges of the new continents and therefore experienced extreme erosion from flood water run-off. This solves the mystery of the huge submarine canyons we find along the edges of the continental shelves. They were once above sea level. The continents were riding much higher than they are today. It took centuries for the continents with their newly formed mountains to sink into the mantle. Today the continents are said to be in ***isostacy*** with the underlying rock, meaning that they are "floating" at the correct level as predicted by their densities. (An ice cube doesn't float on top of the water, but is halfway submerged into the water, at its predicted isostacy level.) It took many centuries for the continents to sink to the level they are today. There were also many post-flood lakes, some larger than the Great Lakes of North America, so not all of the flood water was in the oceans right after the flood. (One of these lakes was in Death Valley.) These factors combined to make the global sea level much lower than it is today. As the continents sank and many post-flood lakes eventually drained into the oceans, the global sea level slowly rose.

Hydroplate Theory agrees that the Bahamas Bank was out of the water for quite some time after the flood, and experienced much erosion during that time. It also agrees that the Bahamas platform has experienced additional growth upwards due to the growth of coral and shelled marine creatures. The top of the platform is thick with organic limestone. Not many places in the world have thick layers of genuine organic limestone.

As the sea level gradually rose, the warm, limey waters around the Bank provided an ideal environment for massive growth of many marine organisms. Coral formations grew quickly upwards, much faster than they do today, and microscopic organisms like foraminiferans also contributed to the limey, organic mass. If you examine beach sand from the Bahamas you can find tiny shells like those shown here.

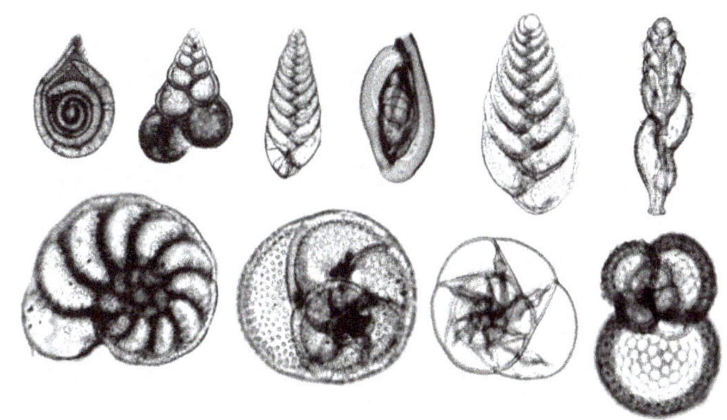

Foraminiferans are tiny (some microscopic) animals that build shells from minerals, usually calcium carbonate.

No discussion of organic limestone is complete without a stop at the White Cliffs of Dover. These famous cliffs are located on England's southern shore, along the English Channel. They are made of the softest type of limestone: white chalk. These White Cliffs are half of a complete landform. The other half lies on the opposite side of the Channel, along the northern coast of France. The opposing cliffs have matching stripes of flint going through them, making it obvious that they were once closer together, if not touching. The Channel was cut long after the chalk was deposited. Germany and Denmark also have some white cliffs, so this chalk formation is quite long.

The chalk in these cliffs is amazingly pure. If you were to look at a sample of this chalk under a microscope, you'd see some really strange shells. Much of the chalk is composed of the exoskeletons of one-celled animals called ***coccolithophores*** *(cock-o-LITH-o-fores)*. This has to be one of the strangest looking creatures on the planet. As we learned on page 6, the living cell is inside this ball of armored plates. It is a photosynthesizing algae, using sunlight to make its own food. Why the animal makes a shell like this is unknown. Coccolithophores are found in all ocean waters and are responsible for a considerable percentage of worldwide production of oxygen. They also put CO_2 into the air and water, according to the chemical equation governing formation of calcium carbonate. The life cycle of coccolithophores is quite short, so the ocean has a steady "rain" of these shells falling to the seafloor. Ocean floors are covered with a mud called "ooze," which has astronomical numbers of these microscopic shells. (In parts of some oceans, foraminiferans and diatoms predominate instead of coccolithophores.)

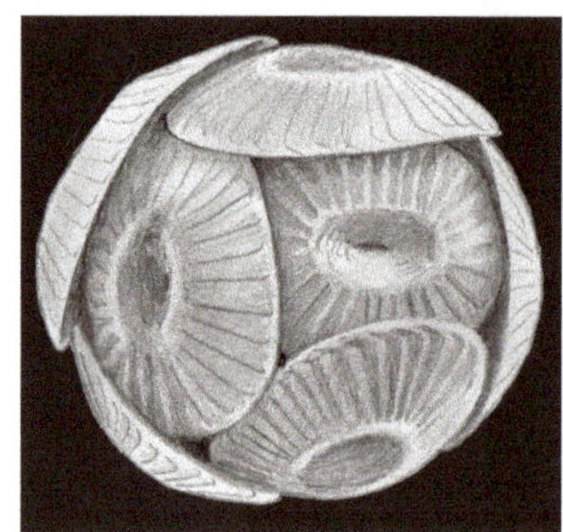

a coccolithophore

To form the large volume of coccolithophore shells we see in the Cliffs of Dover, there would have to have been some major algal blooms over the course of several months near the end of the flood. Single-celled algae can reproduce very quickly if conditions are favorable; the population can grow exponentially. Blooms of various types of algae get so large and dense that they can be seen from satellites. Conditions at the end of the flood would have been ideal for exponential growth of coccolithophores. The water was warm and full of calcium and carbonates for making shells. There could easily have been blooms at frequent intervals for many weeks. As the shells of the coccolithophores piled up on the bottom, they quickly covered the remains of marine creatures that had died recently. Most of the fossils found in chalk are of animals with shells or exoskeletons, such as brachiopods, ammonites and sea urchins. These creatures would have benefited from high levels of calcium and other minerals in the water.

The cliffs are eroding faster now, at 22-32 cm/year.

If all this chalk had built up over millions of years would it be this pure? (The dark stuff you see in the photo is just vegetation clinging to the surface. The chalk itself is pure white.) The coccolithophores that are collecting on the ocean floors today are not forming pure white chalk; thus, the present is not the key to the past when it comes to chalk formation. Judging by what we see happening in the present, the earth will never again produce chalk like this. Unfortunately, we see much erosion happening today. If the cliffs really are 90 million years old, and erosion happened at the rate geologists estimate—5 cm per year— the cliffs should be gone by now and the Channel should be twice as wide.

Now that we've looked at two limestone formations where living organisms really did play an important role in their formation, let's revisit the claim that ALL limestone was formed in warm, shallow seas from the shells of marine creatures. A large part of the eastern Unitied States is made of limestone. This limestone is thick, gray, and has no fossils. To look at it, you'd never guess it was chemically similar to the Bahamas or the Dover cliffs. There aren't any coccolithophores or foraminiferans or corals. Once in a long while a small fossil will turn up, or a chunk of some other type of rock (easily explained by flood debris floating around), but for the most part it's boring gray rock that makes good gravel. There isn't any evidence that suggests an organic origin.

We saw this mountain made of dolomitic limestone on page 3. Clams made all that rock? It's hard to believe a geologist can say that with a straight face. Also, remember that clams and corals don't use $MgCO_3$ in their shells, only $CaCO_3$. This gives us another reason to cancel the clam story. The dolomite likely goes far under the mountain, too, which adds to the unbelievability of the clam story. Hydroplate Theory suggests that dolomite formed at the same time, and in the same way, as limestone. The magnesium could have come from dissolved mantle rock, which is believed to be high in Mg and Fe. Limestone sediments high in Mg made dolomite.

Caves represent another type of limestone that did not come from ancient seas. First a massive layer of inorganic limestone was laid down. Then, a large volume of water was needed to hollow out the basic shape of the cave. (Recently, it has been discovered that in caves of the western USA, sulfuric acid may also have played a significant role.) During the centuries following the flood, water trickling down through the ceiling of the cave brought calcium carbonate with it, and as evaporation happened, the calcium carbonate was deposited as stalactites and stalagmites. The process might have happened quite quickly at first, when water was abundant right after the flood. In some caves today, stalactites are growing very slowly. Assuming that "the present is the key to the past," geologists then calculate that theses formations must be millions of years old. For example, a sign at Carlsbad Caverns in New Mexico used to say the caverns are 260 millions of years old. Then new data began to come in, and the sign was changed to say 8 million years. Then more data came in and the sign was changed to say 2 million years. In 2011, one of the Carlsbad park rangers witnessed a stalactite grow several inches in a few days. The sign was then removed altogether. Tour guides at the caverns now say that the rate of cave formation depends on how much water is present.

Cave formation is one topic that most flood theories agree upon. Besides the initial formation of the limestone, HPT's interpretation of caves is not that different from other flood theories. Everyone agrees that after the flood there was a large volume of mineral-saturated water above and below the surface, allowing for more rapid precipitation than we have today.

Was there limestone, shale and sandstone on the surface of the pre-flood earth? So far, we've not discovered a way to find out. There wouldn't have been any fossiliferous limestone, since fossils appear to be the result of the flood, but we can't rule out limestone being present in certain locations, perhaps provided by God as a raw material that pre-flood people could have used for making buildings. If so, it is likely that most of it was crushed, eroded or dissolved during the Flood. Hydroplate Theory's energetic fountains give the most effective erosional mechanism. Sloshing flood waters might have done some erosion, but not nearly enough to have produced the vast amounts of sediment needed to form our geological record. Tidal waves (even large ones) could not have produced the volume of sedimentary rock visible in the Grand Canyon. We see minimal erosion today along shorelines where waves are constantly crashing on the rocks. Even powerful waterfalls can't erode rock faster than about a meter per year.

This map shows the places around the world that have carbonate rocks, most of which are limestone but could also include dolomite or travertine. Not all of these red areas are solid limestone. Many are patchy and "discontinuous" and have been simplified for clarity. Some of these red areas have several layers of limestone under the surface, going down hundreds of meters. If the "warm, shallow sea" story is correct, all these red places would have to have been seas at least once, and often multiple times. Notice that limestone occurs even in very northern latitudes. Geologists "solved" this problem by proposing that the continents have drifted all over the planet multiple times. This "solution" actually creates more problems than it solves. Discussing these problems is outside the scope of this book. See the bottom of page 39 for suggested reading about the problems with plate tectonics.

There is a famous philosophical principle called "Occam's Razor," attributed to William of Ockham in the 14th century. He said, "Entities must not be multiplied beyond necessity." In simple language, this means the simplest explanation is most likely to be the right one. This idea pre-dates Ockham, and has probably occurred to many people both before and after him. If we apply this principle to limestone, what is the simplest solution to the mysteries of limestone?

Hydroplate Theory offers a mechanism for producing vast quantities of inorganic limestone and sorting them into very flat layers. It has a mechanism for producing a one-time sliding event that moved the continents and formed mountain ranges. It has a relatively simple explanation for metamorphic rock, such as marble, being produced near the surface (lateral compression). It explains why the Mid-Atlantic Ridge looks the way it does. Like other flood theories, it has good answers for the formation of coal, fossils, and caves. As an added bonus, this same theory can shed light on other mysteries that are outside the scope of this book—mysteries related to meteorites, comets, and asteroids, as well as radioactivity in continental granites. A theory that can answer more questions than it raises merits consideration. Like all theories, it will change and grow as time goes on. More data will become available, which might strengthen some of its assertions or serve to adjust others. A new generation of scientists will pick it up and add to it with their own insights and research.

BIBLIOGRAPHY

This book is based on the limestone chapter from *In The Beginning; Compelling Evidence for Creation and the Flood,* by Dr. Walt Brown. ISBN 978-1-878026-09-7. Published by Center for Scientific Creation, Phoenix, AZ. His book is now in its 9th edition and can be read online at creationscience.com or https://hpt.rsr.org/flipbook/ (Also see: www.hydroplate.org.)

Special thanks to Bryan Nickel for the use of diagrams from his video series (Youtube, Rumble) on Hydroplate Theory.

ONLINE RESOURCES (RESEARCH PAPERS, WEBSITES, VIDEO INTERVIEWS)

Types of limestone
https://www.gsa.gov/node/88304?Form_Load=88341
https://geology.com/rocks/limestone.shtml#chemical
https://en.wikipedia.org/wiki/List_of_types_of_limestone
https://en.wikipedia.org/wiki/Oolite
https://en.wikipedia.org/wiki/Aragonite
https://creationscience4kids.com/limestone-travertine-and-tufa/

Dolomite/dolostone
https://en.wikipedia.org/wiki/Dolomite_(mineral)
https://geology.com/rocks/dolomite.shtml (article by Hobart M. King, PhD)

Precipitation of limestone particles in ocean water
Broecker, Wallace & Takahashi, Taro. (1966). Calcium carbonate precipitation on the Great Bahama Banks. Journal of Geophysical Research. 71. 10.1029/JZ071i006p01575. (accessed via this site: https://www.researchgate.net/publication/264699246_Calcium_carbonate_precipitation_on_the_Great_Bahama_Banks)
https://www.geological-digressions.com/tag/great-bahama-bank/
https://www.geological-digressions.com/the-mineralogy-of-carbonates-non-skeletal-grains/

How much limestone there is in earth's crust
P. Falkowski et al., "The Global Carbon Cycle: A Test of Our Knowledge of Earth as a System," Science, Vol. 290, 13 Oct. 2000, p. 293.

Limestone in the Grand Canyon versus modern limestone
https://www.icr.org/article/were-grand-canyon-limestones-deposited-by-calm-pla/
E.D. McKee and R.G. Gutschick, History of the Redwing Limestone in Northern Arizona (Boulder, Colorado, Geological Society of America, Memoir 114, 1969
R.P. Steinen, "On the Diagenesis of Lime Mud: Scanning Electron Microscopic Observations of Subsurface Material from Barbados, W.I.," Journal of Sedimentary Petrology 48
Z. Lasemi and P.A. Sandberg, "Transformation of Aragonite-dominated Lime Muds to Microcrystalline Limestones," Geology 12
R.M. Garrels and F.T. Mackenzie, Evolution of Sedimentary Rocks (New York, W.W. Norton, 1971).
R.L. Folk, "Practical Petrographic Classification of Limestones," American Association of Petroleum Geologists Bulletin, 43 (1959)

Tidal pumping under the crust of Enceladus
https://www.princeton.edu/news/2016/04/12/tidal-forces-explain-how-icy-moon-saturn-keeps-its-tiger-stripes
https://en.wikipedia.org/wiki/Enceladus

Out-salting by supercritical water
Hovland, Martin, et al. "Sub-surface Precipitation of Salt in Supercritical Seawater." *Basin Research*, Vol. 18, 2006, p. 221.
Hannay, J. B. and James Hogarth. "On the Solubility of Solids in Gases." Proceedings of the Royal Society, Vol. 29, 1879, p. 326.

Minerals found in meteorites
http://www.galleries.com/rocks/meteoric.htm
https://sites.wustl.edu/meteoritesite/items/quartz-calcite-magnetite-hematite-micas/
https://en.wikipedia.org/wiki/Carbonaceous_chondrite (also lists specific meteorites that contain amino acids and DNA bases)
https://sites.wustl.edu/meteoritesite/items/chemical-composition-of-meteorites/
https://www.sciencedirect.com/science/article/pii/S0016703714005146 (carbonate minerals in chondritic meteors)

Water in mantle
https://www.livescience.com/46292-hidden-ocean-locked-in-earth-mantle.html
https://physicstoday.scitation.org/doi/abs/10.1063/PT.3.1476?journalCode=pto
https://en.wikipedia.org/wiki/Ringwoodite
https://www.livescience.com/44057-diamond-inclusions-mantle-water-earth.html

Water on Mars
https://www.nasa.gov/press-release/nasa-confirms-evidence-that-liquid-water-flows-on-today-s-mars
https://www.space.com/mars-liquid-water-south-pole-subglacial

Water below the crust of the earth
Wei, Wenbo et al. "Detection of Widespread Fluids in the Tibetan Crust by Magnetotelluric Studies." *Science,* vol. 292, April 27, 2001.
Makovsky, Yizhaq, and Simon Kemperer. "Measuring the Seismic Properties of Tibetan Bright Spots: Evidence for Free Aqueous Fluids in the Tibetan Middle Crust." *Journal of Geophysical Research,* vol. 104, no. B5. May 1999.
https://en.wikipedia.org/wiki/Kola_Superdeep_Borehole
Emmerman, Rolf, and Jorn Lauterjung. "The German Continental Deep Drilling Program KTB: Overview and major results." Journal of Geophysical Research, vol. 102, No. B8, pages 18,179-18,201. 1997
Hyndman, R. D. et al. "The Origin of Electrically Conductive Lower Continental Crust: Saline Water or Graphite?" *Physicis of the Earth and Planetary Interiors,* Vol. 81, 1993, pages 325, 341.

Evidence for the remains of the collapsed chamber
Rychert, Catherine A. "The Slippery Base of a Tectonic Plate." Nature, Vol. 518, 5 February, 2015, pp. 39-40.
Stern, T. A., et al. "A Seismic Reflection Image for the Base of a Tectonic Plate," Nature, Vol., 5 Feb. 2015, 518, pp. 85-88

Properties of SCW
"Sub-surface precipitation of Salts in Supercritical Seawater," Basin Research, Vol. 18, 2006, p.221.

Evaporite formation and salt deposition under the Mediterranean Sea (order of precipitation for limestone, salts)
https://www.cambridge.org/core/journals/european-review/article/uncovering-the-mediterranean-salt-giant-medsalt-scientific-networking-as-incubator-of-crossdisciplinary-research-in-earth-sciences/16535E0F27EA6455EB30166C82EE0B07

Dissolved silica and the necessary ppm to petrify wood
Sigelo, Anne C. "Organic Geochemistry of Silicified Wood, Petrified Forest National Park, Arizona." Geochemica et Cosmochimica Acta, Vol. 42, Sept. 1978, pp. 1397-1405.

Lineaments are tensile fractures
https://www.sciencedirect.com/science/article/abs/pii/0191814182900049
https://docplayer.net/127429013-Lineament-analysis-south-florida-region.html

Charles Lyell and geology
https://creation.com/the-hidden-agenda-of-charles-lyell-was-to-free-science-from-moses

Sedimentary rock layers in eastern North America
https://earthathome.org/hoe/se/rocks-ib/

Coal
https://www.wkms.org/2011-04-07/shark-fossil-found-in-western-kentucky-coal-mine
https://www.smithsonianmag.com/history/the-worlds-largest-fossil-wilderness-30745943/
https://www.palaeontologyonline.com/articles/2011/fossil-focus-coal-swamps/?doing_wp_cron=1682730716.4335780143737792968750
https://www.uky.edu/KGS/coal/coal-mining-geology-coal-balls.php
https://slate.com/technology/2012/11/coal-mine-fossils-paleontology-shows-us-past-climate-change.html
https://www.youtube.com/watch?v=RR7cCYm1nVQ (people going into old mine and finding fossils)
http://allanmccollum.net/amcnet3/reprints/parkerr.html (dinosaur tracks in Utal coal mines)
https://en.wikipedia.org/wiki/Blackhawk_Formation (dino tracks in ceiling of coal mines in this formation)
https://www.youtube.com/watch?v=0aonGWZjKS8 (Making coal in a lab, with Kurt Wise on "Is Genesis History?")

Objection to liquefaction
https://answersresearchjournal.org/noahs-flood/ichnofossils-hydroplate-theory/

Making rocks in a lab
https://www.youtube.com/watch?v=0aonGWZjKS8 (Dr. Kurt Wise, "Is Genesis History?")

Carbon 14 dating of limestone
Syarifuddin, Muhammad Zakir, *Alfian Noor. "Age Determination Of Limestone Rocks Leang- Leang Cave Compiler Through Activity Measurements 14C Using LSC." Marina Chimica Acta, Oktober 2014, Vol. 15 No.2, Jurusan Kimia FMIPA, Universitas, ISSN 1441-2132/

Shale
https://en.wikipedia.org/wiki/Burgess_Shale

Depositional layers (megsequences) across three continents
https://answersresearchjournal.org/sedimentary-record-flooding-sauk/

Planar lamination in sedimentary rocks
https://www.youtube.com/watch?v=wVDxgzMO3mk (Steve Austin's lecture: "The Mudrock Revolution")
https://www.youngearth.com/grand-canyon-nautiloids
Austin, S. A. 2003. "Nautiloid Mass Kill and Burial Event, Redwall Limestone (Lower Mississippian), Grand Canyon Region, Arizona and Nevada." In Proceedings of the Fifth International Conference on Creationism. Ivey, Jr., R. L., ed. Pittsburgh, PA: Creation Science Fellowship, 55-100.
Austin, S. A. 1984. "Rapid Erosion at Mount St. Helens." In Origins.11 (2): 90-98. Geoscience Research Institute, Loma Linda, CA.

Cambrian rock
https://geologictimepics.com/2012/03/28/cambrian-rock/
https://ucmp.berkeley.edu/cambrian/cambrian.php
https://www.nps.gov/articles/000/cambrian-period.htm

Death Valley limestone
https://en.wikipedia.org/wiki/Geology_of_the_Death_Valley_area
https://www.youtube.com/watch?v=4Emw2wqd1Iw&list=PLL26lo9ULg4REruqXORzhQy0a0I-tBgF3&index=7 (Andrew Dunning)
https://www.nps.gov/deva/learn/nature/geologicformations.htm (list of Death Valley statigraphy)
https://commons.wikimedia.org/wiki/File:Striped_Butte_in_Butte_Valley.jpg

Bahamas bank
https://www.scienceandthesea.org/program/201008/great-bahama-bank
https://en.wikipedia.org/wiki/Bahama_Banks
"Geomorphology from Space, Chapter 6: Coastal Landforms. Plate C-16, 'Great Bahama Bank'". geoinfo.amu.edu.pl.
http://geoinfo.amu.edu.pl/wpk/geos/GEO_6/GEO_PLATE_C-16.HTML
https://www.grandbahamamuseum.org/exhibits/natural-history/geology

White cliffs of Dover, Germany and Denmark
https://www.cliffsofdover.com/fossils-of-the-cliffs-of-dover/types-of-fossils-in-the-cliffs-of-dover/
https://en.wikipedia.org/wiki/White_Cliffs_of_Dover https://en.wikipedia.org/wiki/Coccolith
https://phys.org/news/2016-11-england-white-cliffs-dover-eroding.html
https://www.grida.no/resources/3001 (chalk cliff of Germany)
https://creation.com/can-flood-geology-explain-thick-chalk-beds

Varves and Kelvin-Helmholtz instability
Steve Austin https://www.youtube.com/watch?v=wVDxgzMO3mk&t=896s

Carlsbad Cavern age
https://www.youngearth.com/carlsbad-cavern

For more information about problems with Plate Tectonics:

1) *Tectonic Globaloney* by N. Christian Smoot. Xlibris, Corp. (c) 2004. ISBN 978-1413437287.
2) *20 Reasons to Question Plate Tectonics* by Ellis Hughes. Published by THM Books in USA, (c) 2023. ISBN 978-1-7374763-9-9.
3) https://www.davidpratt.info/tecto.htm
4) "Continental Drift, V: Proposed Hypothesis of Earth Tectonics." Meyerhoff, A.A. et al. The University of Chicago Press, 1972. (https://www.journals.uchicago.edu/doi/abs/10.1086/627794)
5) "Against the Hypothesis of Ocean-floor Spreading" by V.V. Beloussov. Tectonophysics 9, 1970. Amsterdam: Elsevier Pub. Co.

www.ingramcontent.com/pod-product-compliance
Lightning Source LLC
LaVergne TN
LVHW060643110826
845147LV00018B/1031
* 9 7 9 8 9 8 6 8 6 3 7 3 3 *